Bismuth Oxyhalides

Synthesis and photocatalytic applications

Online at: https://doi.org/10.1088/978-0-7503-5934-4

Bismuth Oxyhalides

Synthesis and photocatalytic applications

Muhammad Ikram
Department of Physics, Government College University Lahore, Katchery Road Lahore, Lahore 54000, Pakistan

Muhammad Ahsaan Bari
Department of Physics, Government College University Lahore, Katchery Road Lahore, Lahore 54000, Pakistan

IOP Publishing, Bristol, UK

ISBN 978-0-7503-5934-4 (ebook)
ISBN 978-0-7503-5932-0 (print)
ISBN 978-0-7503-5935-1 (myPrint)
ISBN 978-0-7503-5933-7 (mobi)

DOI 10.1088/978-0-7503-5934-4

Version: 20240301

IOP ebooks

British Library Cataloguing-in-Publication Data: A catalogue record for this book is available from the British Library.

Published by IOP Publishing, wholly owned by The Institute of Physics, London

IOP Publishing, No.2 The Distillery, Glassfields, Avon Street, Bristol, BS2 0GR, UK

US Office: IOP Publishing, Inc., 190 North Independence Mall West, Suite 601, Philadelphia, PA 19106, USA

We dedicate this book to our beloved parents and esteemed teachers, whose unwavering support, endless prayers, and unyielding encouragement have been cornerstones of our educational and professional development. We are grateful for their invaluable guidance and wisdom, which shaped our life journey. Their insights may continue to illuminate our path as we navigate the complexities of life.

Contents

Preface

The exponential expansion of industrialization and rapid increase in population have led to the emergence of significant humanitarian concerns in the 21st century, including the global energy crisis and environmental pollution. The use of cost-effective and sustainable energy sources for the generation of power and mitigation of pollutants is widely regarded as the most effective approach to tackle these pressing issues. Consequently, much attention has been directed on the conversion of solar energy into a viable energy source using diverse technologies such as photocatalysis, solar cells, and photoelectrochemical cells. Moreover, the continuously increasing need for fossil fuels, accompanied by the corresponding rise in atmospheric carbon dioxide levels, has created a pressing need for the expeditious advancement of carbon management systems. The pressing environmental and energy challenges necessitate the expeditious development, formulation, and verification of meticulously engineered photocatalytic materials. These catalysts possess the capability to harness light or sunshine, hence facilitating a range of reactions and/or generating energy.

Considering aforementioned problems, this book examines the utilization of bismuth oxyhalide-based photocatalysts for several applications, including the degradation of organic pollutants, the creation of hydrogen and oxygen, and the reduction of carbon dioxide and nitrogen from the atmosphere. It presents a comprehensive examination of bismuth, encompassing its physical and chemical characteristics and a thorough examination of bismuth-based materials, including their various phases and characteristics in relation to photocatalysis, has been undertaken. A concise summary of the synthesis of materials based on bismuth has been presented. It also presents a comprehensive study on the properties and structure of the bismuth oxyhalides as well as a detailed examination of the techniques employed in the synthesis of bismuth oxyhalides has been presented.

Furthermore, this book presents an in-depth examination of the challenges related to the photocatalytic process of bismuth oxyhalides. Additionally, it provides a concise overview of the existing solutions that have been created to address these concerns. The primary emphasis of this work is on the uses of bismuth oxyhalides in environmental, energy, and pollutant removal contexts, namely through the process of photocatalysis. The photocatalytic water splitting process of bismuth oxyhalides to produce hydrogen and oxygen along with its detailed chemical route has been presented. Furthermore, it encompasses a thorough examination of the hydrogen generation method, alongside an in-depth analysis of the reduction mechanisms pertaining to carbon dioxide and nitrogen. The utilization of bismuth oxyhalides for water disinfection through photocatalysis has been examined, alongside an exploration of its potential mechanism. It also includes an examination of the breakdown pathway of pollutants and the removal of antibiotics from wastewater by the application of photocatalysis, specifically utilizing bismuth oxyhalides photocatalysts. The book holds significance for academics in both academia and industry who are engaged in the analysis of photocatalytic applications, as well as energy conversion and generation by using bismuth oxyhalide photocatalysts.

Acknowledgements

In the name of ALLAH (SWT), the creator, the most gracious, the most merciful, and worthy of worship. All praises be to the almighty ALLAH (SWT), who guides us when we are bewildered in the darkness of ignorance and enlighten our ways. We offer countless salutations to the source of knowledge, our Holy Prophet Muhammad (SAW), who is a torch of guidance, assistance, and knowledge for humanity in every aspect of life.

Ahsaan shows great pleasure to express his sincere gratitude to Dr. Muhammad Ikram (Assistant Professor), Department of Physics, Government College University Lahore, for his able guidance, enthusiasm, competent advice, constructive criticism, little castigation, constant encouragement, valuable suggestions, and painstaking supervision for completion of this book. We would like to express our gratitude to our families for their motivation, remembering us in prayer, and especially for the great patience they showed during this journey in the book. Thanks to everyone on our publishing team.

The Authors

Author biographies

Muhammad Ikram

Dr. Muhammad Ikram obtained his Master's degree (M. Phil Physics) from BZU Multan, Pakistan in 2010. He obtained his Ph.D. degree in Physics from the Department of Physics, Government College University (GCU) Lahore through the Pak-US joint project between the Department of Physics, GCU Lahore, Pakistan, and the University of Delaware, USA in 2015. He served as deputy director of Manuscript Science at Punjab textbook board (Pakistan). Later on (2017-to date), Ikram joined the Department of Physics, GC University Lahore as an Assistant Professor of Physics and Principal of Solar Cell Applications Lab. Ikram received the Seal of Excellence Marie Skłodowska Curie Actions Individual Fellowship in 2017 and 2020. In 2021 and 2022, Ikram was included in the 2 % top scientists from Pakistan announced by Stanford University. His research interest involves the synthesis and characterization of inorganic semiconductor nanomaterials, 2D materials for water treatment optoelectronic, and electrocatalytic applications. Ikram has written two international Springer books and one book with IOP Publishing

Muhammad Ahsaan Bari

Muhammad Ahsaan Bari obtained his B.S. (Physics) degree from Bahauddin Zakariya University (Punjab, Pakistan) in 2016 and his Master's degree in Material Physics from Government College University, Lahore, Pakistan, in 2022. Bari did his M.Phil. research at GCU Lahore (Punjab, Pakistan) in the Solar Cell Applications Lab. He is then working as a Research Associate in the same lab on the field of catalytic and photocatalytic applications of nanomaterials.

IOP Publishing

Chapter 1

Bismuth based materials

In the past few decades, attention toward the utilization of materials containing bismuth for various sustainable environmental and energy applications has grown. This heightened emphasis can be attributed mostly to the low toxicity and environmentally friendly nature of these materials. Bismuth-based materials are utilized extensively in electrochemical energy conversion as well as storage devices, showcasing remarkable non-catalytic and catalytic capabilities, along with their applications in the reduction of carbon dioxide, nitrogen fixation, and polluted water treatment. A diverse range of bismuth-based materials, such as oxyhalides, oxides, bismuthates, chalcogenides, and various composites, have been synthesized and studied in order to gain insights into their physicochemical characteristics. This chapter presents a comprehensive examination of the properties shown by individual Bi material systems. Furthermore, a detailed study of the materials based on bismuth is presented, including an exploration of their many phases and the conditions under which they are synthesized. Our research centers mostly on the investigation of bismuth oxyhalides, namely examining their characteristics and conducting structural analysis. Finally, an analysis is presented on the current methods used for synthesizing compounds based on bismuth.

1.1 Bismuth

Bismuth (Bi) is a semimetal with overlapping valence and conduction bands that belongs to group V in the periodic table of elements [1]. Bi is a very rare element in nature, and, as a result, it carries a significant market value. It is present in ores in the form of oxides and sulfides. In particular, bivalent compounds have garnered a substantial amount of interest over the past several decades. This is because bivalent Bi is the heaviest element that is non-toxic among its neighbors on the periodic table. Micro- as well as nanostructured materials based on Bi have garnered a lot of attention recently, attributed to the distinctive characteristics exhibited by these materials, which are not present in their bulk counterparts. In particular, these Bi-

doi:10.1088/978-0-7503-5934-4ch1

based micro- and nanostructured materials have shown advances in their intrinsic material properties, such as their electrical, catalytic, photon absorption, and optical qualities. For instance, Bi–metal nanoparticles (NPs) with high-density have been the subject of extensive research due to their one-of-a-kind pseudo-layered structure, tunable bandgaps, and compositional modifications. These characteristics have resulted in their excellent performance in energy conversion and storage systems, catalysis, antimicrobial drugs, photonics, and optoelectronics [2–7]. Bi has a very low free carrier density compared to its neighboring metals in the periodic table, which ranges between 1018 and 1019 cm^{-3} [8, 9]. Using localized surface plasmon resonance, Bi NPs may also exhibit plasmonic activity (≈ 5–10 nm), which is advantageous for capturing sunlight and reducing the possibility of charge recombination [10]. According to a report, 63% of Bi is used to make cosmetics, pharmaceuticals, and pigments, 26% for casting and galvanizing, 7% for alloys, ammunition, and solders, and 4% for research [11]. Materials based on Bi have found widespread use in the field of photocatalysis due to the large variety of compositional forms available, the one-of-a-kind crystal structure features, and the simplicity of directed modification. As stated previously, elemental Bi shows semimetallic behavior, whereby it can be stimulated to produce photoelectrons or function as an electron trap. When it comes to compounds, the valence of bismuth is predominantly +3. Its 6s orbital is easily capable of hybridization with the orbitals of other elements, which enables the formation of a new hybrid state and, as a result, a reduction in the band gap for photoexcitation as well as the formation of highly dispersive band structures [12]. To date, numerous kinds of photocatalysts based on Bi have been reported in the literature, which include monobasic bismuth compounds such as Bi_2O_3 [13] and Bi_2S_3 [14], binary bismuth oxygenates such as Bi_2WO_6 [15, 16], Bi_2MoO_6 [17], $Bi_2Sn_2O_7$ [18], and $Bi_3Ti_4O_{12}$ [19], bismuthates such as $NaBiO_3$ [20], $AgBiO_3$ [21], and $LiBiO_3$, as well as bismuth composite oxides such as CsAgBiBr [22], PbBiOBr [23], Bi_4VO_8Cl [24], and Bi_4TaO_8Cl [25] (figure 1.1). In particular, bismuth oxyhalide [26] (BiOX, where X represents Cl, Br, or I) has garnered significant attention because of its one-of-a-kind structure, which is made up of alternate layers of $[Bi_2O_2]^{2+}$ and halogen. The indirect band gap of BiOX necessitates the involvement of phonons in the electron drop process, which slightly slows down the rate at which electrons/holes can recombine. Despite the advancements achieved in the realm of photocatalysis using Bi-based materials, their practical use on an industrial scale remains a distant prospect.

1.2 Properties of bismuth

1.2.1 Physicochemical properties

Bi is a brittle metal found in nature as a free metal or in a variety of minerals [28]. Bi crystals have a spiral staircase-like structure because the growth of Bi occurs more quickly on the outside edges than the inside edges. Oxides of Bi are produced when it is subjected to the atmosphere. The layers of bismuth oxide (Bi_2O_3) that are formed have varied thicknesses, and they are layered on the surface of the crystal. As a result, at various wavelengths these layers show varying degrees of light reflection, which

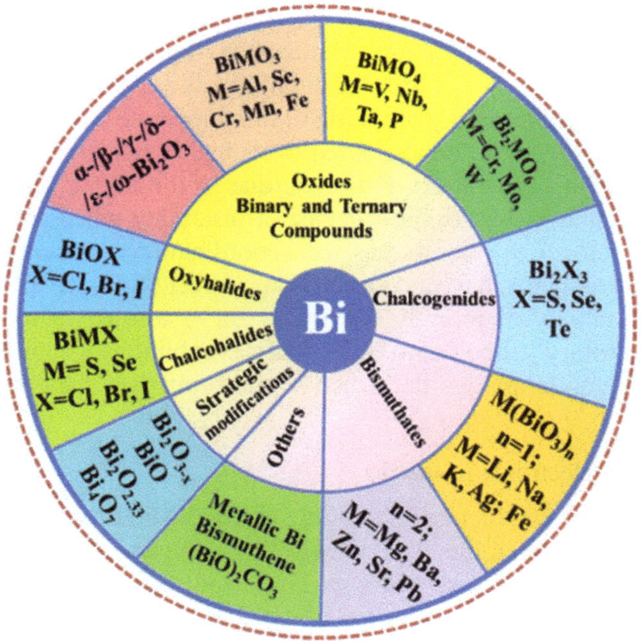

Figure 1.1. Classification of Bi-based materials. (Reproduced with permission from [27]. Copyright 2022 John Wiley and Sons.)

results in the rainbow colors on the bismuth crystal surface and crystal structure. Bi exhibits significantly poor heat conductivity, except for the case of mercury. Bismuth demonstrates a melting point of 271 °C as well as a boiling point of up to 1560 °C. The density of liquid Bi is higher than that of solid Bi, exhibiting similarity to water. It was thought that the periodic table's most stable and heaviest element was Bi. However, recent research has shown that Bi is also mildly radioactive. Earlier studies reported that ^{209}Bi undergoes alpha decay to become ^{205}Ti, accompanied by an astonishing half-life of about 1.9 (±0.2) × 10^{19} years [29–34]. This calculated value is significantly greater, by a factor of one billion, than the present estimated age of the Universe. Bi's toxicity is lower than that of its neighbors, such as antimony, tin, polonium, tellurium, and lead. Bi exhibits two noteworthy oxidation states, namely Bi(V) and Bi(III). Among them, Bi(III) is more prevalent as well as persistent, readily undergoing hydrolysis in aqueous environments. The complexity of the Bi(III) nitrite aqueous solution is worth noting. However, the Bi(III) ion undergoes disintegration into products comprising hydrogen and oxygen, the specific composition of which is contingent upon the pH level. The Bi(III) ion exerts a strong attractive force in the direction of ligands that are composed of nitrogen and oxygen [11]. Upon interaction with thiol-containing ligands such as glutathione and cysteine, the Bi(III) ion undergoes a transformation, leading to the formation of complexes facilitated by thiolate linkages. Several reports have documented the formation of Bi complexes with ligands, including amino polycarboxylate and polyaminopolycarboxylate. These complexes exhibit a range of coordination numbers, varying from 3 to 10. This diversity in coordination numbers indicates the irregular structure of the compounds. For instance, some complexes adopt a tricapped trigonal prism structure with a coordination number of 9, while others exhibit a pyramid structure with a coordination number of 3 [35]. The ability of the Bi(V) ion to operate as a powerful oxidant in solutions based on water can be attributed to its higher reductive potential (2.03 V) when going from Bi(V) to Bi(III). However, due to the higher reductive potential, complexes containing Bi(V) exhibit a degree of instability when present in aqueous solutions. Except for a few complexes with coordination numbers of 6 or higher, several Bi-based complexes have a coordination number of 5, such as Ar_3Bi $(HCO_2)_2$. In light of the stability of recently discovered Bi(V) complexes such as the Bi(V) tropolonato and tri(aryl)tropolonato Bi(V) complex exhibiting a coordination number of 7, it is challenging to analyse the structure and properties of Bi-based complexes.

1.2.2 Optical and electronic properties

Because of their narrower band gap, nanomaterials (NMs) that are based on Bi have a strong reaction to light in the ultraviolet to the visible spectrum. Bi NPs that were monodispersed and generated using a liquid pulsed laser ablation technique demonstrated a larger sensitivity to light from 220 to 600 nm [36], as reported by Verma and co-workers. It was discovered that Bi NPs were produced at a quick rate at the early stage, which lasted for 15–60 min. After that, the number of NPs produced steadily decreased as time increased, lasting for around 60–90 min. This is because when the ablation process's duration grows, light scattering increases slightly. Additionally, utilizing the liquid phase epitaxy method and the *N*-methyl-2-pyrrolidone solvent, Bi-based quantum dots (QDs) were synthesized [37]. A comprehensive range of absorbance from 200 to 600 nm was observed in the prepared Bi-based QDs. This supported Verma's findings that Bi QDs are well-suited for UV–visible photoresponse applications [36]. The Tyndall effect shows that the produced Bi QDs are rapidly re-diffused in *N*-methyl-2-pyrrolidone and are conveniently stable without aggregation when stored under ambient conditions for two weeks. Furthermore, the application of Bi-based NMs extends to other areas such as mid-infrared lasers, which can serve as optical modulators operating at a wavelength of 2 μm [38]. These NMs also find utility in Q-switched solid-state lasers operating at 1.3 μm, as well as all-optical phase modulators and intensity modulators in conjunction with 900 nm lasers [39]. This phenomenon occurs due to the comparatively low light absorption capacity of NMs in the near-infrared region. Bi-based NMs can be used in an extensive applications, such as biomedical and UV–visible-light-based photoelectric devices, thanks to their significant light absorption capabilities throughout a wide band gap.

The intriguing properties of Bi NPs have recently garnered greater interest in electronics owing to their narrow band gap, improved mobility of carriers, and enhanced stability [40, 41]. Xing and co-workers used a liquid phase epitaxy process to manufacture QDs based on Bi. Their response to light was evaluated using a photoelectrochemical set-up [42]. The electrode used in this study was a bare glass electrode based on indium tin oxide (ITO). This electrode was exposed to light with varying wavelengths between 300 and 800 nm while being immersed in a 0.1 M KOH solution. Nevertheless, despite the laser's high power density, the bare electrode only displayed a minor photoresponsive response. The outstanding observation is that the QDs of Bi exhibit different switching behavior, specifically between an 'On' and 'Off' state, throughout a range of wavelength spanning from 350 to 400 nm. Moreover, it has been shown that there is a positive correlation between the density of the laser and the photocurrent density. This indicates that the photoresponse signal is mostly generated by the samples of Bi rather than the final product. A reduction in photoresponse is observed when Bi QDs are exposed to a 350 nm laser, suggesting that the strong density of laser power could generate a large amount of pairs of electrons and holes and, consequently, photocurrent. This is true even though the laser power density is elevated, causing the QDs based on Bi to have a continually rising photocurrent density. These extraordinary electrical properties of NPs based on Bi would pave the way for understanding the new structure of field effect transistors based on Bi NMs, which would result in higher activity.

1.3 Bismuth based materials

1.3.1 Bismuth oxide and its phases

The phase structures of Bi_2O_3 play a significant role in determining its chemical and physical properties. Epsilon (ε-Bi_2O_3, tetragonal), omega (ω-Bi_2O_3, triclinic), delta (δ-Bi_2O_3, face-centered cubic), gamma (γ-Bi_2O_3, body-centered cubic), beta (β-Bi_2O_3, tetragonal), and alpha (α-Bi_2O_3, monoclinic) are the reported phases of Bi_2O_3 [43, 44]. In particular, α-Bi_2O_3 having a monoclinic structure and δ-Bi_2O_3, having a face-centered cubic structure, are the stable phases at high and low temperature, respectively. The tetragonal-structured β-Bi_2O_3 phase and the body-centered cubic γ-Bi_2O_3 phase are both metastable phases of the compound. At temperatures over a particular threshold, distinct crystalline phases can metamorphose into one other. At room temperature, Bi_2O_3 is in the α phase, and its melting point is 824 °C. If it is heated to 729 °C, it can change into the δ phase. The δ phase will precipitate throughout the cooling process at temperatures of 650 °C and 639 °C, respectively, to produce the β phase and γ phase (figure 1.2(a)) [45]. The oxygen vacancies are distributed unevenly throughout the δ-Bi_2O_3 structure. The structure contains oxygen ions that have great mobility, which demonstrates that the oxygen ions have high conductivity [46]. Both β-Bi_2O_3 and δ-Bi_2O_3 have structures that have an arrangement of oxygen vacancies that are quite similar to one another. The energy of the band gap in Bi_2O_3 may range anywhere between 2.1 and 3.96 eV [47, 48], with β-Bi_2O_3 and α-Bi_2O_3 having a band gap energy (E_g) of 2.58 and 2.85 eV, respectively. The influence that the E_g has on the photocatalytic activity (PCA) of Bi_2O_3 could be elaborated by the orbital energy levels that it has (figure 1.2(b)). Bi 6p levels make up the conduction band of Bi_2O_3, while Bi 6s, along with O 2p hybrid orbitals, make up the valance band. Both of these sublevels are present in all of the phases. The manipulation of the lone pair Bi 6s orbital leads to a reduction in the E_g and an increase in charge mobility. This manipulation causes the Bi 6s orbital to become distorted, resulting in a noticeable overlap between the Bi 6s orbital and the O 2p

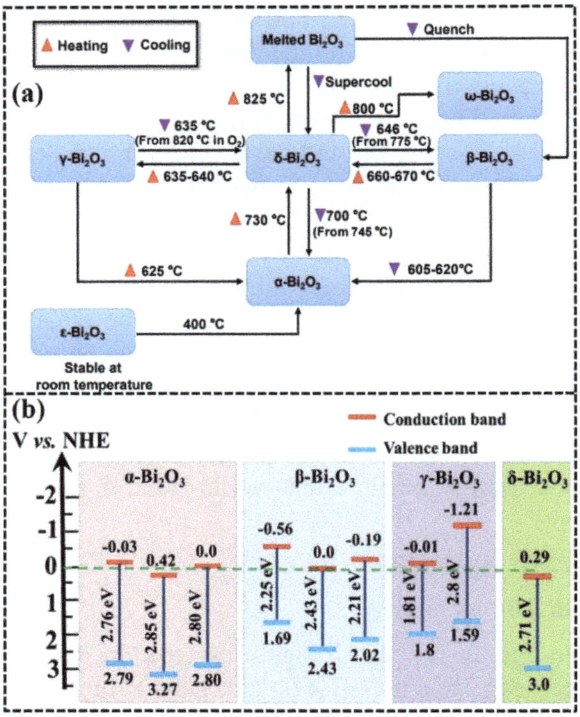

Figure 1.2. (a) Conditions of phase transitions between Bi_2O_3 phases. (Adapted with permission from [60]. Copyright 2020 The Royal Society of Chemistry.) (b) Energy level diagram of Bi_2O_3 at different phases. (Adapted with permission from [61]. Copyright 2021 The Royal Society of Chemistry.)

orbital [49, 50]. Strong repulsion is exerted on other bonds by the isolated pair of $6s^2$ electrons in the Bi_2O_3 molecule. Within the valance band of a Bi-semiconductor, the Bi 6s electronic orbitals as well as the O 2p hybridizations can be found in a position with a lower energy level. This effect causes an excessive amount of distortion in the lattices as well as a polar electric field, which ultimately results to an elevation in charge separation. The valance band shift to higher energy levels is the cause of the reduced E_g value of β-Bi_2O_3, which may be compared to the band gap values of other polymorphs. The existence of Bi 6s lone pair electrons in the valence band causes the electron mobility to rise after being exposed to radiation, and the *c*-axis-oriented tetragonal structure of β-Bi_2O_3 exhibits better electron transport. The lone pair also inhibits the recombinations of pairs of electron and hole and serves as a channel for the photoexcited movement of electrons [51]. Under ambient conditions, Bi_2O_3 is typically recognized as a p-type semiconductor photocatalyst. As a result, it has an outstanding capacity to conduct electrons during photocatalysis [52]. Bi_2O_3 crystals have various desirable characteristics including a high refractive index, exceptional dielectric property, remarkable photoluminescence, strong oxygen ion conductivity, and great photoconductivity performance, contingent upon the phase structure. These properties are made possible by the many different crystal types and the vast band gap width that can be adjusted. As a consequence of this, Bi_2O_3 is utilized extensively in a variety of sectors, including high refractive index glasses, high-temperature superconducting materials, microelectronic components, sensors, optoelectronic materials, and catalysis [53].

The compound Bi_2O_3 exhibits exceptional PCA upon exposure to visible light (VL), making it highly promising for numerous applications in photocatalysis. The intrinsic shortcomings of Bi_2O_3 prevent it from being used as a sole catalyst to break down organic contaminants. These defects include a high rate of loading photogenerated electron holes, a lack of photocorrosion susceptibility, and structural stability [54, 55]. As a result, the primary objectives of the ongoing study are to strengthen the structural stability of Bi_2O_3 to make it more resistant to light corrosion and to improve its reaction to VL. Further investigation is required in the areas of recycling, loading properties, interfacial properties, and the integration of different composite modification approaches, in addition to the PCA mechanisms of the Bi_2O_3 heterojunction complex. Bi_2O_3 also includes monoclinic dibismuth tetraoxide (m-Bi_2O_4). Due to its unique electron band structure and outstanding ability to absorb sunlight, this compound has lately attracted the attention of scientists working in photocatalysis [56–59]. Wang and colleagues developed submicrorods of m-Bi_2O_4 with a low E_g (2.0 eV). These submicrorods have remarkable PCA in the VL range, enabling the breakdown of organic contaminants and the eradication of germs. The submicrorods exhibit a maximum absorption wavelength of 620 nm [58]. In this context, m-Bi_2O_4 has the potential to be a stable and original photocatalyst with a visible-light response. If this proves to be the case, it will have significant application possibilities in energy conversion and environmental restoration. Nevertheless, the present state of research pertaining to morphological regulation, synthesis methods, and modification of the electrical structure of m-Bi_2O_4 remains inadequate. Therefore, future investigations should prioritize these aforementioned features in order to enhance and maximize the potential advantages of m-Bi_2O_4.

1.3.2 Bismuth tungstate

The compound bismuth tungstate (Bi_2WO_6) is highly sought after due to its diverse range of uses in producing electronic equipment. This material is adaptable and could be utilized for different applications, including thermistors, humidity detectors, optical sensors, and capacitors [62–64]. This particular substance is a member of the Aurivillius phase ($n = 1$) family, which has the general formula $Bi_2A_{n-1}B_nO_{3n+1}$ (with A = Na, Bi, Pb, Ba, Sr, Ca, and K and B = Fe, W, Mo, Ta, Nb, and Ti) [65]. The structure's composition consists of alternating layers of $[Bi_2O_2]^{2+}$ ions resembling fluorine, as well as $[WO_4]^{2-}$ ions. These layers are joined to create n perovskite-like units, which have the formula $A_{n-1}B_nO_{3n+1}$. The discovery of its photocatalytic properties under visual irradiation, namely for water oxidation to generate dioxygen, was initially reported in 1999 [66]. These properties were then utilized in 2003 by Ye and colleagues for the degradation of organic molecules (acetaldehyde and chloroform) [67]. Since then, its PCA has been put to extensive use to break down a broad variety of compounds, whether they are in gaseous or aqueous form, and even microbes when exposed to VL [68, 69]. In the presence of VL, the crystal and electronic structure of Bi_2WO_6 are, in fact, well-suited for photocatalysis. The compound Bi2WO_6 exhibits orthorhombic symmetry, with its structure consisting of alternating layers of Bi_2O_2 sheets and WO_6 octahedra and these layers are interconnected through their vertices. This arrangement of alternately charged (−) and (+) sheets results in the generation of an electric field that is perpendicular to the plane in which the sheets are arranged (001). This electric field enhances the mobility of photoexcited charges inside the material as well as making it easier for molecules to adsorb onto the surface [70, 71]. Because of its remarkable ferroelectricity, piezoelectricity, catalytic capabilities, and nonlinear dielectric susceptibility, Bi_2WO_6 has been the subject of a large amount of investigation ever since its discovery by Knight [72]. Since Bi_2WO_6 has an E_g of 2.8 eV, it is capable of absorbing longer wavelengths of light, making it an excellent candidate for use in applications using solar energy. Despite continued study into Bi_2WO_6 and its many derivatives, the topic is still substantially undeveloped [71].

1.3.3 Bismuth titanate

Titanate-based photocatalysts have garnered notable attention over the past several decades as a result of their availability and their low impact on the surrounding ecosystem [73–75]. Bismuth titanate ($Bi_4Ti_3O_{12}$) is classified as a perovskite oxide belonging to the Aurivillius phase with a wide E_g of approximately 2.9–3.0 eV. In the form of an orthorhombic crystal structure, the $Bi_4Ti_3O_{12}$ includes layers that alternate between $(Bi_2O_2)^{2+}$ and $(Bi_2Ti_3O_{10})^{2-}$ [74, 76, 77]. On the other hand, it is impossible to successfully separate the photogenerated carriers of pure $Bi_4Ti_3O_{12}$, which leads to poor performance of pure $Bi_4Ti_3O_{12}$ for hydrogen evolution [75]. It should come as no surprise that a significant amount of focus and energy should be directed into effectively preventing the recombination of photoinduced electrons and holes. It is essential to modify the $Bi_4Ti_3O_{12}$ compound. For instance, coupling with semiconductors based on silver not only helps to preserve the broad band gap energy of $Bi_4Ti_3O_{12}$, but also encourages the transmission of electron–hole pairs that have been photogenerated [78].

1.3.4 Bismuth vanadate

Bismuth vanadate, $BiVO_4$, is an n-type semiconductor that is both non-toxic and inexpensive. It also possesses exceptional photostability and chemical properties. $BiVO_4$'s ferroelastic, acousto-optical, photocatalytic, and ionic conductivity properties have helped it achieve widespread technical applications in several fields over the past few years [79]. These features are highly dependent on the $BiVO_4$'s electronic and crystal structures. In general, $BiVO_4$ may be found in the form of three different polymorphs, each of which has a distinct crystal structure. These three polymorphs are known as dreyerite, pucherite, and clinobisvanite. The pucherite polymorph is the mineral that may be found naturally occurring as $BiVO_4$, and it contains orthorhombic crystal structure. Crystallographically,

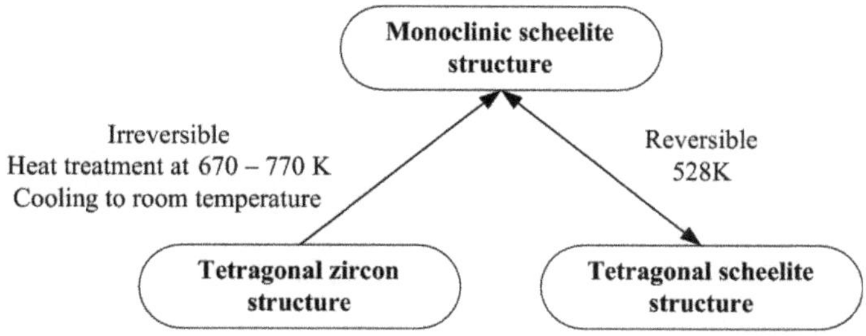

Figure 1.3. Phase transitions of $BiVO_4$. (Reproduced by permission from [81]. Copyright 2019 Springer.)

dreyerite is characterized by a tetragonal zircon structure, whereas clinobisvanite may take on either a tetragonal or monoclinic scheelite arrangement. In the tetragonal structure of zircon, as well as the tetragonal and monoclinic $BiVO_4$ structures, four O atoms are responsible for stabilizing the vanadium ion. In comparison, eight O atoms are responsible for coordinating the Bi ion. However, the O atoms around Bi ion come from eight separate VO_4 tetrahedral units in the scheelite formations. In contrast, the oxygen atoms surrounding the Bi ion come from just six VO_4 units in the tetragonal zircon structure. The primary distinction between monoclinic scheelite as well as tetragonal scheelite structure is that the monoclinic scheelite structure exhibits a greater degree of distortion in the local surroundings of Bi ions and vanadium ions.

Additionally, the presence of distinct crystal structures of $BiVO_4$ is strongly influenced by temperature, and it is usual for there to be transitions from one crystal structure to another during the process. For example, the shift from tetragonal zircon structure to monoclinic scheelite structure in $BiVO_4$ occurs when the material is heated to temperatures between 670 K and 770 K and then cooled to room temperature. However, this transformation cannot be reversed once it has taken place. The tetragonal structure of scheelite, on the other hand, may change into the monoclinic structure of scheelite at a temperature of 528 K (figure 1.3) [80]. Additionally, it has been shown that mechanical grinding performed under ambient conditions can change the tetragonal structure of $BiVO_4$ into a monoclinic structure.

The research findings indicate that the monoclinic scheelite phase, among the several crystal structures of $BiVO_4$, exhibits the highest PCA for the process of water oxidation when exposed to VL. This was linked to the distortion of the local environment of BiO_8 as well as VO_4, respectively. This distortion arises due to the phenomenon of overlap or hybridization between a Bi ion occupying the 6s orbital and an oxygen atom occupying the 2p orbital located in the uppermost region of the valence band. This phenomenon facilitates enhanced electron mobility induced by photoexcited electrons and restricts the recombination of pairs of electrons and holes [82]. Furthermore, distortion facilitates a more efficient and expedited transfer of photoinduced electrons to the vanadium 3d orbitals located in the conduction band of monoclinic scheelite $BiVO_4$. This is what accounts for the material's reduced E_g, which ultimately results in enhanced optical qualities. More specifically, the E_g of the monoclinic structure of $BiVO_4$ is around 2.4 eV, whereas the E_g of the tetragonal structure is within the range of 2.9–3.1 eV [83, 84]. As a result of the remarkable PCA of monoclinic $BiVO_4$, it has found widespread use in the process of photodegrading a large variety of organic contaminants [85–88]. Because of the favorable location of its valence band, it is also a material of interest for the production of a photoanode, which is necessary for the evolution of hydrogen during the process of water splitting [89–92]. In addition to this, $BiVO_4$ has been utilized in the production of photoelectrochemical sensors [93].

1.3.5 Bismuth molybdate

Bismuth molybdate (BMO), a photocatalyst that is driven by VL, has attracted a substantial amount of interest in recent years due to the fact that it is non-toxic, efficient, and responsive to VL [94]. There are three different phases that bismuth molybdate may exist in, and they are γ-BMO, β-BMO, and α-BMO. The standard chemical formula for BMO is $Bi_2O_3{\cdot}nMoO_3$, where $n = 3, 2, 1$. This formula corresponds to three distinct phases, referred to as γ-Bi_2MoO_6, β-$Bi_2Mo_2O_9$, and α-$Bi_2Mo_3O_{12}$, in that order [95]. α-BMO possesses a scheelite structure, and within that structure are three separate MoO_4 units. The crystal structure of β-BMO is fluorite with MoO_4 tetrahedra [96, 97]. An Aurivillius structure is present in the γ-BMO phase, and it consists of perovskite layers of $(MoO_4)^{2+}$ sandwiched between layers of $(Bi_2O_2)^{2+}$ [98, 99]. Because it possesses the highest photocatalytic performance, γ-Bi_2MoO_6 has emerged as a front-runner in the photocatalysis field. In addition, the edge of Bi_2MoO_6's VL sensitivity may be extended to a wavelength of around 500 nm [100, 101]. Its conduction band consists of Bi 6p and Mo 4p orbitals, while its valence band is composed of O 2p orbitals [102]. Bi_2MoO_6 has gotten very little attention as a method for generating H_2 by water splitting despite the many advantages it offers. Zhang *et al* used a hydrothermal anion exchange technique to generate porous nanoflake $BiMO_x$ (M = Mo, V, and W) photoanodes for their photocathode experiment [103]. Note that the formation of Bi_2MoO_6 occurred when WO_4^{2-} in bismuth tungstate was substituted with MoO_4^{2-}. The researchers employed a porous nanoflake Bi_2MoO_6 material as a photocatalyst in water splitting. Consequently, they achieved a noteworthy outcome of 120 μA cm^{-2}, ranked among the highest recorded results for Bi_2MoO_6. There is still a significant amount of work that has to be done to split water using Bi_2MoO_6 as a photocatalyst.

1.3.6 Bismuth chalcogenides

Bi-based chalcogenides [104] are n-type semiconductors with a small E_g and strong ionic and photonic conductivity. Bi-based chalcogenides include Bi_2S_3, Bi_2Se_3, and Bi_2Te_3 [105]. In general, Bi_2S_3 can be described as having a layer-based anisotropic orthorhombic crystal structure. The interlayers are held together by weak van der Waals contacts, and the material has a direct E_g that ranges from 1.3 to 1.7 eV owing to the different micro- and nanostructures [106]. Bi_2Se_3, bismuth selenide, exhibits a layered and laminated structure, with each layer having a thickness of approximately 0.96 nm. The layers composed of five atoms arranged in a specific order, namely Se–BiSe–Bi–Se, forming covalent bonds along the *z*-axis. Consequently, the interlayer forces of attraction are relatively weak, facilitating the disintegration of Bi_2Se_3 into discrete nanosheets. Since the carrier mobility of charges of this compound is high and the E_g is low, the compound has significant potential applications in photoelectrochemical and thermoelectrical devices, as well as in photocatalysis and optical recording equipment. Bi_2Te_3, also known as bismuth telluride, has an elevated melting temperature of 585 °C, an E_g of 0.15 eV, and forms a trigonal crystal structure. When brought to room temperature and combined with selenium or antimony, this combination undergoes a transformation that enables it to function as an efficient thermoelectric compound. Because of these unusual properties, bismuth telluride is currently being used in the refrigeration industry as well as in the generation of power through the use of thermoelectricity [11].

1.3.7 Bismuth oxyhalides (BiOX)

The crystal structure of BiOX is classified as PbFCl type, exhibiting D4h symmetry and belonging to the P4/nmm space group. Additionally, BiOX is a member of the tetragonal system. The crystal structure of BiOX has a $[Bi_2O_2]^{2+}$ layer that is interlaced with double halogen ions; this gives the BiOX material a high degree of anisotropy (figure 1.4) [107, 108]. The [X–Bi–O–Bi–X] layers of BiOX exhibit elongation along the (001) crystallographic direction and this elongation allows for the production of an internal electric field (IEF) in the same (001) direction, facilitated by the layers possessing a positive charges and the adjacent [X] layers carrying a negative charges [45]. As a result of the unique coordination that occurs around the Bi core in each layer of [X–Bi–O–Bi], BiOX possesses an asymmetric decahedral geometry [109]. Covalent bonds are responsible for the interactions in the $[Bi_2O_2]^{2+}$ layer. In contrast, van der Waals forces, also known as nonbonding

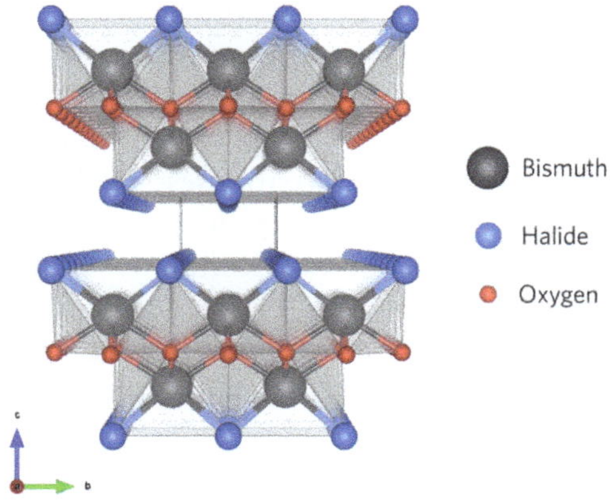

Figure 1.4. Framework of BiOX systems with stoichiometric X–Bi–O–Bi–X bilayers stacked along the *c*-axis. (Reproduced from [116]. CC BY 4.0.)

interactions, are responsible for the interactions in the [X] layer, which is superimposed on top of the $[Bi_2O_2]^{2+}$ layer [110]. Van der Waals interaction between weak layers and covalent bonding between strong layers can both cause the development of layered structures. The photogenerated pairs of electron and hole that are created as a result of photoexcitation can be readily separated when subjected to the action of an IEF, which is beneficial to the photocatalytic process [111, 112].

Regarding the BiOX crystal, the valence band maximum is established by combining the O 2p state and the X np state, where X represents the elements I, Br, and Cl with corresponding values of n being 5, 4, and 3, respectively. On the other hand, the conduction band minimum is mostly influenced by the Bi 6p state [113–115]. Because of the greatly increased contribution of X ns states, the band gap will be shrunk when the atomic number of X grows, and the dispersive quality of the band energy level will become more apparent. The redox potentials and E_g values of BiOX can be significantly influenced by the composition of its layered structure, such as the atomic numbers of X. BiOF is the least studied member of the BiOX series of semiconductors owing to its large band gap and direct band gap structure. All three of these band gap transitions, BiOI, BiOBr, and BiOCl, are indirect. Regarding photocatalytic applications, the three semiconductors (BiOI, BiOBr, and BiOCl) each have their own unique set of benefits and drawbacks. With an E_g of about 3.2 eV, BiOCl has a low absorption capacity for VL in sunlight and a high threshold for UV radiation. BiOBr has an E_g of around 2.9 eV, and its response range to VL is not very high. BiOI has the narrowest E_g of any known material, indicating its exceptionally high ability to absorb VL. Nevertheless, the PCA of BiOI is constrained by its restricted redox capacity, which arises from challenges associated with the positioning of its redox potential in both the conduction band and the valence band. This restricts the photocatalytic use of BiOI. Therefore, additional modification of the photocatalytic materials containing BiOX is required in order to attain increased PCA and broaden its practical applicability.

BiOBr is a p-type semiconductor material that is also an efficient photocatalyst because it possesses outstanding optical features, is chemically stable, and has improved PCA. It possesses a tetragonal crystal structure characterized by layered arrangements, demonstrates excellent performance in the VL range and exhibits efficient charges separation capabilities, enabling the degradation of contaminants. BiOBr microspheres were produced by Zhang *et al* utilizing a ethylene glycol (EG) assisted solvothermal technique [117]. Because BiOBr has a small E_g, it demonstrated 3.5 times and 2.5 times stronger PCA than TiO_2 when subjected to VL and UV irradiation, respectively. In recent years, the use of BiOBr has been observed under visible irradiation due to its comparatively smaller E_g than BiOCl.

Additionally a p-type semiconductor material, BiOCl possesses outstanding optical, catalytic, and electrical characteristics, in addition to a good ion conductivity. This molecule has the shape of a tetragonal crystal and comprises ions of chlorine (Cl), Bi, and oxygen stacked one on top of the other in the order Cl–Bi–O–

Bi–Cl. In the tetragonal structure of the pyramidal phase, the central position is occupied by the Bi atom. The configuration of the atoms makes it such that the four Bi–O atoms and the four Bi–Cl atoms are placed in a manner that is diametrically opposed to one another. It has been observed that the BiOCl is effective in the region of UV where the theoretical E_g values levels between 2.8 and 2.9 eV and the experimentally observed values of E_g values are between 3.1 and 3.5 eV [109]. Zhang *et al* conducted a synthesis of BiOCl using the hydrolysis method. Their study revealed, for the first time, that BiOCl exhibited a higher level of efficacy compared to TiO_2 (P25, Degussa) in terms of its photocatalytic breakdown capabilities for methyl orange (MO) dye [109].

Another molecule, known as BiOI, has a tetragonal structure that is layered and is made up of an alternate slab of iodine (I) as well as $(Bi_2O_2)^{2+}$ ion. This is one of the captivating NPs for the PCA of the compound under VL because of its tapered E_g between 1.6 and 1.9 eV, which provided the compound with higher activity under VL. The effectiveness of hole–electron separation in VL can be attributed to the tetragonal structure of BiOI, which exhibits a robust intralayer bonding and moderate van der Waals force of attraction [118].

1.4 Synthesis of bismuth-based photocatalysts

The hydrothermal approach may be used to produce Bi-based photocatalysts [119, 120]. This process includes the growth of crystals developed in solutions under high pressure and at a temperature of around 300 °C. The autoclave is the typical setting for the execution of this procedure. Therefore, it is feasible to identify certain qualities before the creation of the catalyst. The shape and material crystallinity may be varied by adjusting several reaction parameters, including the quantity of Bi source that was utilized [119]. The development of Bi-based catalysts may also be accomplished by using chemical reduction, which can be aided by either electricity or light. Because the polyol contains many hydroxyl groups, this process is often mediated by the use of polyol so that the characteristics of the Bi-based NPs can be improved [121]. The sol–gel method is a further approach that, compared to the other approaches, may be executed at a significantly lower temperature [122, 123]. It is a straightforward method that results in products that have a high level of purity and also makes it possible to modulate morphologies. In general, Bi-based NPs may be designed in a variety of nanostructures, and the performance of these NPs is dependent on their morphology as well as their size [124]. The microemulsion method is another synthesis process that may modulate features such as surface area, homogeneity, geometry, size, and morphology [125]. Specifically, one may change the size and form of the particles by modifying the proportion of water to surfactant. The inclusion of surfactants reduces the surface tension and prevents agglomeration [125, 126]. There is also the possibility of developing Bi-based catalysts by the use of microwave irradiation, which is a procedure that is rapid, straightforward, and requires minimal reaction time [127]. It also makes it possible for morphologies to change, and the products it yields often have a limited distribution and tiny crystalline sizes [128]. Other approaches that have been reported include an ultrasonic route, direct heating, the solvent evaporation method, pulse laser ablation, solvothermal synthesis, solution evaporation, spin coating, ion thermal synthesis, and a modified version of the Pechini method [129, 130]. At times it may be necessary to combine these techniques to achieve the appropriate characteristics in the Bi-based NPs. For instance, the synthesis of nanoplates of Bi_2O_3 was accomplished by employing a precipitation process aided by ultrasound [131]. In general, the application of the photocatalysts provides clues regarding the method of synthesis that should be utilized.

References

[1] Xu K, Wang L, Xu X, Dou S X, Hao W and Du Y 2019 Two dimensional bismuth-based layered materials for energy-related applications *Energy Storage Mater.* **19** 446–63

[2] Xia L, Fu W, Zhuang P, Cao Y, Chee M O L, Dong P, Ye M and Shen J 2020 Engineering abundant edge sites of bismuth nanosheets toward superior ambient electrocatalytic nitrogen reduction via topotactic transformation *ACS Sustain. Chem. Eng.* **8** 2735–41

[3] Shen K, Zhang Z, Wang S, Ru Q, Zhao L, Sun L, Hou X and Chen F 2020 Cucumber-shaped construction combining bismuth nanoparticles with carbon nanofiber networks as a binder-free and freestanding anode for Li-ion batteries *Energy Fuels* **34** 8987–92

[4] Miola M, De Jong B C A and Pescarmona P P 2020 An efficient method to prepare supported bismuth nanoparticles as highly selective electrocatalyst for the conversion of CO_2 into formate *Chem. Commun.* **56** 14992–5

[5] Wang R, Lai T P, Gao P, Zhang H, Ho P L, Woo P C Y, Ma G, Kao R Y T, Li H and Sun H 2018 Bismuth antimicrobial drugs serve as broad-spectrum metallo-β-lactamase inhibitors *Nat. Commun.* **9** 1 12

[6] Nishikubo R, Kanda H, García-Benito I, Molina-Ontoria A, Pozzi G, Asiri A M, Nazeeruddin M K and Saeki A 2020 Optoelectronic and energy level exploration of bismuth and antimony-based materials for lead-free solar cells *Chem. Mater.* **32** 6416–24

[7] Cao J, Zhang Z, Li X and Peng M 2019 Abnormal NIR photoemission from bismuth doped germanophosphate photonic glasses *J. Mater. Chem.* C **7** 3218–25

[8] Huber T E, Brower T, Johnson S D, Belk J H and Hunt J H 2017 Photocurrent in bismuth junctions with graphene http://arxiv.org/abs/1709.05408

[9] Gallo C F, Chandrasekhar B S and Sutter P H 1963 Transport properties of bismuth single crystals *J. Appl. Phys.* **34** 144–52

[10] Subramanyam P, Deepa M, Raavi S S K, Misawa H, Biju V and Subrahmanyam C 2020 A photoanode with plasmonic nanoparticles of earth abundant bismuth for photoelectro-chemical reactions *Nanoscale Adv.* **2** 5591–9

[11] Sivasubramanian P D, Chang J H, Nagendran S, Dong C-D, Shkir M and Kumar M 2022 A review on bismuth-based nanocomposites for energy and environmental applications *Chemosphere.* **307** 135652

[12] Zhang L, Li Y, Li Q, Fan J, Carabineiro S A C and Lv K 2021 Recent advances on bismuth-based photocatalysts: strategies and mechanisms *Chem. Eng. J.* **419** 129484

[13] Wang L *et al* 2020 Promoted photocharge separation in 2D lateral epitaxial heterostructure for visible-light-driven CO_2 photoreduction *Adv. Mater.* **32** 2004311

[14] Chen X, Li Q, Li J, Chen J and Jia H 2020 Modulating charge separation via *in situ* hydrothermal assembly of low content Bi_2S_3 into UiO-66 for efficient photothermocatalytic CO_2 reduction *Appl. Catal.* B **270** 118915

[15] Lou Z, Lu C, Li X, Wu Q, Li J, Wen L, Dai Y, Huang B and Li B 2021 Constructing surface plasmon resonance on Bi_2WO_6 to boost high-selective CO_2 reduction for methane *ACS Nano.* **15** 3529–39
[16] Li Y Y, Fan J S, Tan R Q, Yao H C, Peng Y, Liu Q C and Li Z J 2020 Selective photocatalytic reduction of CO_2 to CH_4 modulated by chloride modification on Bi_2WO_6 nanosheets *ACS Appl. Mater. Interfaces* **12** 54507–16
[17] Di J *et al* 2019 Atomically-thin Bi_2MoO_6 nanosheets with vacancy pairs for improved photocatalytic CO_2 reduction *Nano Energy.* **61** 54–9
[18] Zhang Y, Di J, Qian X, Ji M, Tian Z, Ye L, Zhao J, Yin S, Li H and Xia J 2021 Oxygen vacancies in $Bi_2Sn_2O_7$ quantum dots to trigger efficient photocatalytic nitrogen reduction *Appl. Catal.* B **299** 120680
[19] Liu L, Huang H, Chen F, Yu H, Tian N, Zhang Y and Zhang T 2020 Cooperation of oxygen vacancies and 2D ultrathin structure promoting CO_2 photoreduction performance of $Bi_4Ti_3O_{12}$ *Sci. Bull.* **65** 934–43
[20] Wu Y, Zhao X, Huang S, Li Y, Zhang X, Zeng G, Niu L, Ling Y and Zhang Y 2021 Facile construction of 2D g-C_3N_4 supported nanoflower-like $NaBiO_3$ with direct Z-scheme heterojunctions and insight into its photocatalytic degradation of tetracycline *J. Hazard. Mater.* **414** 125547
[21] Gong J, Lee C S, Kim E J, Kim J H, Lee W and Chang Y S 2017 Self-generation of reactive oxygen species on crystalline $AgBiO_3$ for the oxidative remediation of organic pollutants *ACS Appl. Mater. Interfaces* **9** 28426–32
[22] Zhou L, Xu Y F, Chen B X, Kuang D-B and Su C Y 2018 Synthesis and photocatalytic application of stable lead-free $Cs_2AgBiBr_6$ perovskite nanocrystals *Small.* **14** 1703762
[23] Wang B *et al* 2020 Revealing the role of oxygen vacancies in bimetallic $PbBiO_2Br$ atomic layers for boosting photocatalytic CO_2 conversion *Appl. Catal.* B **277** 119170
[24] You F, Wei J, Cheng Y, Wen Z, Ding C, Guo Y and Wang K 2020 A sensitive and stable visible-light-driven photoelectrochemical aptasensor for determination of oxytetracycline in tomato samples *J. Hazard. Mater.* **398** 122944
[25] Li S, Wang C, Li D, Xing Y, Zhang X and Liu Y 2022 Bi_4TaO_8Cl/Bi heterojunction enables high-selectivity photothermal catalytic conversion of CO_2–H_2O flow to liquid alcohol *Chem. Eng. J.* **435** 135133
[26] Di J *et al* 2020 Strain-engineering of $Bi_{12}O_{17}Br_2$ nanotubes for boosting photocatalytic CO_2 reduction *ACS Mater. Lett.* **2** 1025–32
[27] Adhikari S, Mandal S and Kim D H 2023 Recent development strategies for bismuth-driven materials in sustainable energy systems and environmental restoration *Small.* **19** 2206003
[28] El-Sayed N Z 2006 Physical characteristics of thermally evaporated bismuth thin films *Vacuum.* **80** 860–3
[29] Halime Z, Lachkar M and Boitrel B 2009 Coordination of bismuth and lead in porphyrins: towards an *in situ* generator for α-radiotherapy? *Biochimie.* **91** 1318–20
[30] Hermanne A, Tárkányi F, Takács S, Szücs Z, Shubin Y N and Dityuk A I 2005 Experimental study of the cross-sections of α-particle induced reactions on ^{209}Bi *Appl. Radiat. Isot.* **63** 1–9
[31] Howell R C, Schweitzer A D, Casadevall A and Dadachova E A 2008 Chemosorption of radiometals of interest to nuclear medicine by synthetic melanins *Nucl. Med. Biol.* **35** 353–7

[32] Kokorian J, Engelen J B C, De Vries J, Nazeer H, Woldering L A and Abelmann L 2014 Ultra-flat bismuth films for diamagnetic levitation by template-stripping *Thin Solid Films* **550** 298–304

[33] Mazarov P, Melnikov A, Wernhardt R and Wieck A D 2008 Long-life bismuth liquid metal ion source for focussed ion beam micromachining application *Appl. Surf. Sci.* **254** 7401–4

[34] Tavares O A P, Medeiros E L and Terranova M L 2005 Alpha decay half-life of bismuth isotopes *J. Phys. G: Nucl. Part. Phys.* **31** 129–39

[35] Yang N and Sun H 2007 Biocoordination chemistry of bismuth: recent advances *Coord. Chem. Rev.* **251** 2354–66

[36] Verma R K, Kumar K and Rai S B 2013 Near infrared induced optical heating in laser ablated Bi quantum dots *J. Colloid Interface Sci.* **390** 11–6

[37] Su X, Wang Y, Zhang B, Zhang H, Yang K, Wang R and He J 2019 Bismuth quantum dots as an optical saturable absorber for a 1.3 μm Q-switched solid-state laser *Appl. Opt.* **58** 1621–5

[38] Chu H, Pan Z, Wang X, Zhao S, Li G, Cai H and Li D 2020 Passively Q-switched Tm: $CaLu_{0.1}Gd_{0.9}AlO_4$ laser at 2 μm with hematite nanosheets as the saturable absorber *Opt. Express* **28** 16893

[39] Wang Y, Huang W, Zhao J, Huang H, Wang C, Zhang F, Liu J, Li J, Zhang M and Zhang H 2019 A bismuthene-based multifunctional all-optical phase and intensity modulator enabled by photothermal effect *J. Mater. Chem.* C **7** 871–8

[40] Ersan F, Kecik D, Özçelik V O, Kadioglu Y, Aktürk O Ü, Durgun E, Aktürk E and Ciraci S 2019 Two-dimensional pnictogens: a review of recent progresses and future research directions *Appl. Phys. Rev.* **6** 021308

[41] Yang Z, Wu Z, Lyu Y and Hao J 2019 Centimeter-scale growth of two-dimensional layered high-mobility bismuth films by pulsed laser deposition *InfoMat.* **1** 98–107

[42] Xing C *et al* 2018 Ultrasmall bismuth quantum dots: facile liquid-phase exfoliation, characterization, and application in high-performance UV–vis photodetector *ACS Photonics.* **5** 621–9

[43] Cornei N, Tancret N, Abraham F and Mentré O 2006 New ε-Bi_2O_3 metastable polymorph *Inorg. Chem.* **45** 4886–8

[44] Brezesinski K, Ostermann R, Hartmann P, Perlich J and Brezesinski T 2010 Exceptional photocatalytic activity of ordered mesoporous β-Bi_2O_3 thin films and electrospun nanofiber mats *Chem. Mater.* **22** 3079–85

[45] Cabot A, Marsal A, Arbiol J and Morante J R 2004 Bi_2O_3 as a selective sensing material for NO detection *Sensors Actuators* B **99** 74–89

[46] Jiang N, Wachsman E D and Jung S H 2002 A higher conductivity Bi_2O_3-based electrolyte *Solid State Ionics.* **150** 347–53

[47] Leontie L, Caraman M, Delibaş M and Rusu G I 2001 Optical properties of bismuth trioxide thin films *Mater. Res. Bull.* **36** 1629–37

[48] Walsh A, Watson G W, Payne D J, Edgell R G, Guo J, Glans P A, Learmonth T and Smith K E 2006 Electronic structure of the α and δ phases of Bi_2O_3: a combined *ab initio* and x-ray spectroscopy study *Phys. Rev.* B **73** 235104

[49] Li M, Huang H, Yu S, Tian N and Zhang Y 2018 Facet, junction and electric field engineering of bismuth-based materials for photocatalysis *ChemCatChem.* **10** 4477–96

[50] Hameed A, Aslam M, Ismail I M I, Salah N and Fornasiero P 2015 Sunlight induced formation of surface Bi_2O_{4-x}–Bi_2O_3 nanocomposite during the photocatalytic mineralization of 2-chloro and 2-nitrophenol *Appl. Catal.* B **163** 444–51

[51] Kim M W, Joshi B, Samuel E, Kim K, Il Kim Y, Kim T G, Swihart M T and Yoon S S 2018 Highly nanotextured β-Bi_2O_3 pillars by electrostatic spray deposition as photoanodes for solar water splitting *J. Alloys Compd.* **764** 881–9

[52] Xu M, Yang J, Sun C, Liu L, Cui Y and Liang B 2020 Performance enhancement strategies of Bi-based photocatalysts: a review on recent progress *Chem. Eng. J.* **389** 124402

[53] Qiu Y, Liu D, Yang J and Yang S 2006 Controlled synthesis of bismuth oxide nanowires by an oxidative metal vapor transport deposition technique *Adv. Mater.* **18** 2604–8

[54] Zhang L, Wang W, Yang J, Chen Z, Zhang W, Zhou L and Liu S 2006 Sonochemical synthesis of nanocrystallite Bi_2O_3 as a visible-light-driven photocatalyst *Appl. Catal.* A **308** 105–10

[55] Huang L, Li G, Yan T, Zheng J and Li L 2011 Uncovering the structural stabilities of the functional bismuth containing oxides: a case study of α-Bi_2O_3 nanoparticles in aqueous solutions *New J. Chem.* **35** 197–203

[56] Kinomura N and Kumada N 1995 Preparation of bismuth oxides with mixed valence from hydrated sodium bismuth oxide *Mater. Res. Bull.* **30** 129–34

[57] Kumada N, Kinomura N, Woodward P M and Sleight A W 1995 Crystal structure of Bi_2O_4 with β-Sb_2O_4-type structure *J. Solid State Chem.* **116** 281–5

[58] Wang W, Chen X, Liu G, Shen Z, Xia D, Wong P K and Yu J C 2015 Monoclinic dibismuth tetraoxide: a new visible-light-driven photocatalyst for environmental remediation *Appl. Catal.* B **176–7** 444–53

[59] Xia D, Wang W, Yin R, Jiang Z, An T, Li G, Zhao H and Wong P K 2017 Enhanced photocatalytic inactivation of *Escherichia coli* by a novel Z-scheme g-C_3N_4/m-Bi_2O_4 hybrid photocatalyst under visible light: the role of reactive oxygen species *Appl. Catal.* B **214** 23–33

[60] Wang S, Wang L and Huang W 2020 Bismuth-based photocatalysts for solar energy conversion *J. Mater. Chem.* A **8** 24307–52

[61] Zahid A H and Han Q 2021 A review on the preparation, microstructure, and photocatalytic performance of Bi_2O_3 in polymorphs *Nanoscale.* **13** 17687–724

[62] Zhao G, Hao S, Xing Y, Wang Y, Wang Y, Xu K and Xu X 2019 Preparation of low-dimensional bismuth tungstate (Bi_2WO_6) photocatalyst by electrospinning *Phys. Status Solidi Appl. Mater. Sci.* **216** 1900035

[63] Zhang K, Wang J, Jiang W, Yao W, Yang H and Zhu Y 2018 Self-assembled perylene diimide based supramolecular heterojunction with Bi_2WO_6 for efficient visible-light-driven photocatalysis *Appl. Catal.* B **232** 175–81

[64] Ait Ahsaine H, BaQais A, Arab M, Bakiz B and Benlhachemi A 2022 Synthesis and electrocatalytic activity of bismuth tungstate Bi_2WO_6 for rhodamine B electro-oxidation *Catalysts.* **12** 1335

[65] Li Y, Liu J and Huang X 2008 Synthesis and visible-light photocatalytic property of Bi_2WO_6 hierarchical octahedron-like structures *Nanoscale Res. Lett.* **3** 365–71

[66] Kudo A and Hijii S 1999 H_2 or O_2 evolution from aqueous solutions on layered oxide photocatalysts consisting of Bi^{3+} with $6s^2$ configuration and d^0 transition metal ions *Chem. Lett.* **28** 1103–4

[67] Tang J, Zou Z and Ye J 2004 Photocatalytic decomposition of organic contaminants by Bi_2WO_6 under visible light irradiation *Catal. Lett.* **92** 53–6

[68] Ren J, Wang W, Zhang L, Chang J and Hu S 2009 Photocatalytic inactivation of bacteria by photocatalyst Bi_2WO_6 under visible light *Catal. Commun.* **10** 1940–3

[69] Fu H, Pan C, Yao W and Zhu Y 2005 Visible-light-induced degradation of rhodamine B by nanosized Bi_2WO_6 *J. Phys. Chem.* B **109** 22432–9

[70] Liu Z, Wu B, Niu J, Huang X and Zhu Y 2014 Solvothermal synthesis of BiOBr thin film and its photocatalytic performance *Appl. Surf. Sci.* **288** 369–72

[71] Ait Ahsaine H, Ezahri M, Benlhachemi A, Bakiz B, Villain S, Guinneton F and Gavarri J R 2016 Novel Lu-doped Bi_2WO_6 nanosheets: synthesis, growth mechanisms and enhanced photocatalytic activity under UV-light irradiation *Ceram. Int.* **42** 8552–8

[72] Knight K S 1992 The crystal structure of russellite; a re-determination using neutron powder diffraction of synthetic Bi_2WO_6 *Mineral. Mag.* **56** 399–409

[73] Chen J, Tang Z, Bai Y and Zhao S 2016 Multiferroic and magnetoelectric properties of $BiFeO_3/Bi_4Ti_3O_{12}$ bilayer composite films *J. Alloys Compd.* **675** 257–65

[74] Liu Y, Zhang M, Li L and Zhang X 2014 One-dimensional visible-light-driven bifunctional photocatalysts based on $Bi_4Ti_3O_{12}$ nanofiber frameworks and Bi_2XO_6 (X = Mo, W) nanosheets *Appl. Catal.* B **160–1** 757–66

[75] Odling G, Chatzisymeon E and Robertson N 2018 Sequential ionic layer adsorption and reaction (SILAR) deposition of $Bi_4Ti_3O_{12}$ on TiO_2: an enhanced and stable photocatalytic system for water purification *Catal. Sci. Technol.* **8** 829–39

[76] Chen Z, Jiang H, Jin W and Shi C 2016 Enhanced photocatalytic performance over $Bi_4Ti_3O_{12}$ nanosheets with controllable size and exposed {001} facets for rhodamine B degradation *Appl. Catal.* B **180** 698–706

[77] Wei W, Dai Y and Huang B 2009 First-principles characterization of Bi-based photocatalysts: $Bi_{12}TiO_{20}$, $Bi_2Ti_2O_7$, and $Bi_4Ti_3O_{12}$ *J. Phys. Chem.* C **113** 5658–63

[78] Liu C, Xu J, Niu J, Chen M and Zhou Y 2020 Direct Z-scheme $Ag_3PO_4/Bi_4Ti_3O_{12}$ heterojunction with enhanced photocatalytic performance for sulfamethoxazole degradation *Sep. Purif. Technol.* **241** 116622

[79] Tan H L, Amal R and Ng Y H 2017 Alternative strategies in improving the photocatalytic and photoelectrochemical activities of visible light-driven $BiVO_4$: a review *J. Mater. Chem.* A **5** 16498–521

[80] Kudo A, Omori K and Kato H 1999 A novel aqueous process for preparation of crystal form-controlled and highly crystalline $BiVO_4$ powder from layered vanadates at room temperature and its photocatalytic and photophysical properties *J. Am. Chem. Soc.* **121** 11459–67

[81] Trinh D T T, Khanitchaidecha W, Channei D and Nakaruk A 2019 Synthesis, characterization and environmental applications of bismuth vanadate *Res. Chem. Intermed.* **45** 5217–59

[82] Cooper J K, Gul S, Toma F M, Chen L, Glans P A, Guo J, Ager J W, Yano J and Sharp I D 2014 Electronic structure of monoclinic $BiVO_4$ *Chem. Mater.* **26** 5365–73

[83] Tolod K R, Hernández S and Russo N 2017 Recent advances in the $BiVO_4$ photocatalyst for sun-driven water oxidation: top-performing photoanodes and scale-up challenges *Catalysts.* **7** 13

[84] Zhang A, Zhang J, Cui N, Tie X, An Y and Li L 2009 Effects of pH on hydrothermal synthesis and characterization of visible-light-driven $BiVO_4$ photocatalyst *J. Mol. Catal.* A **304** 28–32

[85] Dunkle S S, Helmich R J and Suslick K S 2009 $BiVO_4$ as a visible-light photocatalyst prepared by ultrasonic spray pyrolysis *J. Phys. Chem.* C **113** 11980–3
[86] Dong L, Guo S, Zhu S, Xu D, Zhang L, Huo M and Yang X 2011 Sunlight responsive $BiVO_4$ photocatalyst: effects of pH on L-cysteine-assisted hydrothermal treatment and enhanced degradation of ofloxacin *Catal. Commun.* **16** 250–4
[87] Soltani T, Tayyebi A and Lee B K 2018 Enhanced photoelectrochemical (PEC) and photocatalytic properties of visible-light reduced graphene-oxide/bismuth vanadate *Appl. Surf. Sci.* **448** 465–73
[88] Liu X, Guo Z, Zhou L, Yang J, Cao H, Xiong M, Xie Y and Jia G 2019 Hierarchical biomimetic $BiVO_4$ for the treatment of pharmaceutical wastewater in visible-light photocatalytic ozonation *Chemosphere.* **222** 38–45
[89] Sinclair T S, Gray H B and Müller A M 2018 Photoelectrochemical performance of $BiVO_4$ photoanodes integrated with [NiFe]-layered double hydroxide nanocatalysts *Eur. J. Inorg. Chem.* **2018** 1060–7
[90] Vo T G, Chiu J M, Chiang C Y and Tai Y 2017 Solvent-engineering assisted synthesis and characterization of $BiVO_4$ photoanode for boosting the efficiency of photoelectrochemical water splitting *Sol. Energy Mater. Sol. Cells* **166** 212–21
[91] Han H S, Shin S, Kim D H, Park I J, Kim J S, Huang P S, Lee J K, Cho I S and Zheng X 2018 Boosting the solar water oxidation performance of a $BiVO_4$ photoanode by crystallographic orientation control *Energy Environ. Sci.* **11** 1299–306
[92] Jian J *et al* 2019 Embedding laser generated nanocrystals in $BiVO_4$ photoanode for efficient photoelectrochemical water splitting *Nat. Commun.* **10** 9
[93] Ribeiro F W P, Moraes F C, Pereira E C, Marken F and Mascaro L H 2015 New application for the $BiVO_4$ photoanode: a photoelectroanalytical sensor for nitrite *Electrochem. Commun.* **61** 1–4
[94] Guo J, Shi L, Zhao J, Wang Y, Tang K, Zhang W, Xie C and Yuan X 2018 Enhanced visible-light photocatalytic activity of Bi_2MoO_6 nanoplates with heterogeneous Bi_2MoO_{6-x}@Bi_2MoO_6 core–shell structure *Appl. Catal.* B **224** 692–704
[95] Reilly L M, Sankar G and Catlow C R A 1999 Following the formation of γ-phase Bi_2MoO_6 catalyst by *in situ* XRD/XAS and thermogravimetric techniques *J. Solid State Chem.* **148** 178–85
[96] Chen H Y and Sleight A W 1986 Crystal structure of $Bi_2Mo_2O_9$: a selective oxidation catalyst *J. Solid State Chem.* **63** 70–5
[97] van den Elzen A F and Rieck G D 1973 The crystal structure of $Bi_2(MoO_4)_3$ *Acta Crystallogr.* B **29** 2433–36
[98] Teller R G, Brazdil J F, Grasselli R K and Jorgensen J D 1984 The structure of gamma-bismuth molybdate, Bi_2MoO_6, by powder neutron diffraction *Acta Crystallogr.* C **40** 2001–5
[99] Ricote J, Pardo L, Castro A and Millán P 2001 Study of the process of mechanochemical activation to obtain Aurivillius oxides with $n = 1$ *J. Solid State Chem.* **160** 54–61
[100] Ma Y, Jia Y, Wang L, Yang M, Bi Y and Qi Y 2016 Exfoliated thin Bi_2MoO_6 nanosheets supported on WO_3 electrode for enhanced photoelectrochemical water splitting *Appl. Surf. Sci.* **390** 399–405
[101] Lai K, Wei W, Zhu Y, Guo M, Dai Y and Huang B 2012 Effects of oxygen vacancy and N-doping on the electronic and photocatalytic properties of Bi_2MO_6 (M = Mo, W) *J. Solid State Chem.* **187** 103–8

[102] Shimodaira Y, Kato H, Kobayashi H and Kudo A 2006 Photophysical properties and pbotocatalytic activities of bismuth molybdates under visible light irradiation *J. Phys. Chem.* B **110** 17790–7
[103] Zhang J, Wang T, Chang X, Li A and Gong J 2016 Fabrication of porous nanoflake BiMOX (M = W, V, and Mo) photoanodes via hydrothermal anion exchange *Chem. Sci.* **7** 6381–6
[104] Deng J and Zhao Z Y 2018 Electronic structure and optical properties of bismuth chalcogenides Bi_2Q_3 (Q = O, S, Se, Te) by first-principles calculations *Comput. Mater. Sci.* **142** 312–9
[105] Gupta S, Singh R, Anoop M D, Mathur D, Ray K, Kothari S L, Awasthi K and Kumar M 2019 Optical studies on bismuth chalcogenides *Mater. Today Proc.* **10** 142–50
[106] Messalea K A *et al* 2020 Two-step synthesis of large-area 2D Bi_2S_3 nanosheets featuring high in-plane anisotropy *Adv. Mater. Interfaces* **7** 2001131
[107] Cheng H, Huang B and Dai Y 2014 Engineering BiOX (X = Cl, Br, I) nanostructures for highly efficient photocatalytic applications *Nanoscale.* **6** 2009–26
[108] Bhachu D S *et al* 2016 Bismuth oxyhalides: synthesis, structure and photoelectrochemical activity *Chem. Sci.* **7** 4832–41
[109] Zhang K L, Liu C M, Huang F Q, Zheng C and Wang W D 2006 Study of the electronic structure and photocatalytic activity of the BiOCl photocatalyst *Appl. Catal.* B **68** 125–9
[110] Li J, Yu Y and Zhang L 2014 Bismuth oxyhalide nanomaterials: layered structures meet photocatalysis *Nanoscale* **6** 8473–88
[111] Guan M, Xiao C, Zhang J, Fan S, An R, Cheng Q, Xie J, Zhou M, Ye B and Xie Y 2013 Vacancy associates promoting solar-driven photocatalytic activity of ultrathin bismuth oxychloride nanosheets *J. Am. Chem. Soc.* **135** 10411–7
[112] Vikrant K, Kim K H and Deep A 2019 Photocatalytic mineralization of hydrogen sulfide as a dual-phase technique for hydrogen production and environmental remediation *Appl. Catal.* B **259** 118025
[113] Zhao L, Zhang X, Fan C, Liang Z and Han P 2012 First-principles study on the structural, electronic and optical properties of BiOX (X = Cl, Br, I) crystals *Physica* B **407** 3364–70
[114] Zhang H, Liu L and Zhou Z 2012 First-principles studies on facet-dependent photocatalytic properties of bismuth oxyhalides (BiOXs) *RSC Adv.* **2** 9224–9
[115] Wang W, Yang W, Chen R, Duan X, Tian Y, Zeng D and Shan B 2012 Investigation of band offsets of interface BiOCl:Bi_2WO_6: a first-principles study *Phys. Chem. Chem. Phys.* **14** 2450–4
[116] Ganose A M, Cuff M, Butler K T, Walsh A and Scanlon D O 2016 Interplay of orbital and relativistic effects in bismuth oxyhalides: BiOF, BiOCl, BiOBr, and BiOI *Chem. Mater.* **28** 1980–4
[117] Zhang X, Ai Z, Jia F and Zhang L 2008 Generalized one-pot synthesis, characterization, and photocatalytic activity of hierarchical BiOX (X = Cl, Br, I) nanoplate microspheres *J. Phys. Chem.* C **112** 747–53
[118] Liu Q C, Ma D K, Hu Y Y, Zeng Y W and Huang S M 2013 Various bismuth oxyiodide hierarchical architectures: alcohothermal-controlled synthesis, photocatalytic activities, and adsorption capabilities for phosphate in water *ACS Appl. Mater. Interfaces* **5** 11927–34
[119] Hayashi H and Hakuta Y 2010 Hydrothermal synthesis of metal oxide nanoparticles in supercritical water *Materials* **3** 3794–817

[120] Kim H Y, Shin J, Jang I C and Ju Y W 2019 Hydrothermal synthesis of three-dimensional perovskite $NiMnO_3$ oxide and application in supercapacitor electrode *Energies* **13** 36
[121] Wang W Z, Poudel B, Ma Y and Ren Z F 2006 Shape control of single crystalline bismuth nanostructures *J. Phys. Chem.* B **110** 25702–6
[122] Zhang Q, Sando D and Nagarajan V 2016 Chemical route derived bismuth ferrite thin films and nanomaterials *J. Mater. Chem.* C **4** 4092–124
[123] Owens G J, Singh R K, Foroutan F, Alqaysi M, Han C M, Mahapatra C, Kim H W and Knowles J C 2016 Sol–gel based materials for biomedical applications *Prog. Mater Sci.* **77** 1–79
[124] Pan C and Zhu Y 2011 Size-controlled synthesis of $BiPO_4$ nanocrystals for enhanced photocatalytic performance *J. Mater. Chem.* **21** 4235–41
[125] Malik M A, Wani M Y and Hashim M A 2012 Microemulsion method: a novel route to synthesize organic and inorganic nanomaterials. 1st nano update *Arab. J. Chem.* **5** 397–417
[126] Das N, Majumdar R, Sen A and Maiti H S 2007 Nanosized bismuth ferrite powder prepared through sonochemical and microemulsion techniques *Mater. Lett.* **61** 2100–4
[127] Liao X H, Wang H, Zhu J J and Chen H Y 2001 Preparation of Bi_2S_3 nanorods by microwave irradiation *Mater. Res. Bull.* **36** 2339–46
[128] Pradhan S, Das R, Bhar R, Bandyopadhyay R and Pramanik P 2017 A simple fast microwave-assisted synthesis of thermoelectric bismuth telluride nanoparticles from homogeneous reaction-mixture *J. Nanopart. Res.* **19** 69
[129] Yesuraj J and Suthanthiraraj S A 2018 DNA-mediated sonochemical synthesis and characterization of octahedron-like bismuth molybdate as an active electrode material for supercapacitors *J. Mater. Sci., Mater. Electron.* **29** 5862–72
[130] Ashfold M N R, Claeyssens F, Fuge G M and Henley S J 2004 Pulsed laser ablation and deposition of thin films *Chem. Soc. Rev.* **33** 23–31
[131] Sánchez-Martínez D, Juárez-Ramírez I, Torres-Martínez L M and De León-Abarte I 2016 Photocatalytic properties of Bi_2O_3 powders obtained by an ultrasound-assisted precipitation method *Ceram. Int.* **42** 2013–20

Chapter 2

Overview and synthesis route of BiOX

The effectiveness of BiOX photocatalysts in terms of photocatalytic activity is primarily influenced by factors such as their crystalline structure, shape, specific surface area, and size. To date, a range of BiOX species, such as sea urchins, hollow balls, nanoflakes, flowers, spheres, and others, have been successfully synthesized for the purpose of photocatalysis. This chapter presents an overview of various synthesis methods employed for the production of distinct BiOX species. These methods include hydrothermal synthesis, solvothermal synthesis, hydrolysis, ultrasonic/ microwave-assisted synthesis, the two-phase method, the precipitation method, and physical synthesis. Furthermore, a comprehensive examination of each approach and its impact on the size, shape, and surface area of BiOX is examined thoroughly. Lastly, we have summarized highly recommended articles describing their synthesis conditions, the precursor materials employed, and the resulting morphologies.

2.1 Hydrothermal method

The hydrothermal approach is frequently employed to synthesize BiOX photocatalysts owing to its ability to achieve exact manipulation of the morphology, crystallinity, and particle size of the final product, even while operating at lower temperatures [1]. The effect of different hydrothermal reaction conditions on the physicochemical characteristics of BiOX and its photocatalytic activity (PCA) has been studied extensively. These parameters include pH, precursor concentration, surfactant, temperature, and reaction time. In hydrothermal procedure, the precursor, such as the Bi source of $NaBiO_3{\cdot}2H_2O$ or $Bi(NO_3)_3{\cdot}5H_2O$, and the source of halogen, which can be ionic liquids, or HX, KX, NaX, are dissolved in water and then placed into an autoclave that is lined with polytetrafluoroethylene (PTFE). Using a hydrothermal treatment of KBr and $Bi(NO_3)_3{\cdot}5H_2O$ precursors in acetic acid at a temperature of 150 °C for a duration of 18 h, Jiang *et al* effectively prepared BiOBr resembling a flake, which exhibited good PCA towards the destruction of

doi:10.1088/978-0-7503-5934-4ch2

methyl orange (MO) [2]. The best oxidation caused by photocatalytic effectiveness on gaseous Hg^0 under ultraviolet radiation was demonstrated in this particular field by Zhang *et al* for the BiOCl nanosheets produced in an alkaline solvent [3]. This has been attributed to the nanosheets' high specific surface area and the crystal BiOCl's advantageous growth direction. In a quite similar process, flakes of BiOBr were produced using a hydrothermal technique [4]. Under the influence of visible light (VL), the sample produced with a pH value of 9 showed the greatest capacity for ciprofloxacin (CIP) elimination through the process of degradation. In addition, Intaphong and colleagues [5] discovered that the value of pH had a significant impact on the crystal structure as well as the morphology of BiOBr, and the PCA of the compound. When the pH value = 8, the BiOBr microflowers provided the best photodegradation performance for the breakdown of rhodamine B (RhB). This phenomenon might be due to the favorable characteristics of BiOBr microflowers, which facilitate the uniform adsorption of RhB molecules. Feng *et al* [6] employed a hydrothermal approach to manufacture BiOBr, and they adjusted the pH value to alter the internal stress of the BiOBr nanosheets (NSs) that they created. The alteration of the symmetry and band structure of BiOBr, induced by its low strain energy, led to an enhanced rate of charge separation and an accelerated decolorization rate of RhB and MO dye.

A number of different surfactants were incorporated into a hydrothermal process to exert control over the microscopic structure of the BiOX. In this regard, Dou *et al* [7] used $EO_{106}PO_{70}EO_{106}$ (F127) as a surfactant while synthesizing BiOI photocatalysts, utilizing glacial acetic acid as the medium. The quick photoexcited carriers separation was helped by the creation of O_2 thanks to F127-BiOI, which possesses a large specific surface area and a notable augmentation in the number of hydroxyl groups. Zhou and colleagues [8] produced BiOCl NSs successfully by employing dulcitol ($C_6H_{14}O_6$) as the surfactant. Changing the solution pH allowed for customization of the thickness of the BiOCl NSs. The BiOCl synthesized at a pH = 4 contains a thinner layer and showed the maximum PCA towards the breakdown of RhB and tetracycline hydrochloride (TC-HCl). This phenomenon occurred owing to the significantly elevated concentration of oxygen vacancies coupled with the exposure of the (001) crystal plane. Li *et al* [9] were able to successfully construct tetragonal BiOCl NSs with exposed (001) facets by making use of cationic polyacrylamide (C-PAM) as the source of surfactant and chlorine. The BiOCl sample that was produced at 150 °C for 12 h showed the greatest capacity for MO removal through the process of degradation. Cai *et al* [10] produced BiOCl NSs of varying diameters by modifying the amount of xylitol surfactants used in the process. The sample that was synthesized with 0.1 g of xylitol exhibited a concentration of oxygen vacancies and grain boundaries, an unusual polycrystalline structure, and a small band gap. This sample successfully degraded 98% RhB in 20 min when exposed to VL. Therefore, the choice of surfactant is of the utmost importance if one wishes to increase the PCA of BiOX. Therefore, the composition, size, crystalline phase, and morphology of the semiconductors may be altered by manipulating the thermodynamic and kinetic conditions of the process. The versatility of this approach is demonstrated by the fact that a wide variety of acceptable solvents may be utilized in the synthesis process. Furthermore, the selection of solvents not only affects the morphological attributes of BiOXs but also enables precise manipulation of particle size throughout crystal formation.

2.2 Solvothermal method

In contrast to hydrothermal processes, which make use of a solution of water, solvothermal reactions include the consumption of organic solvents. The impact of the solvent is crucial, regardless of the reaction circumstances. When making BiOX, a variety of alcohols are utilized. Ethylene glycol (EG) is the most popular and commonly utilized of these many alcohols. For instance, Zhang *et al* first created three-dimensional BiOX microspheres (MSs) by making use of a solvothermal method with EG (figures 2.1(a)–(f)). He produced BiOX powder (X = I, Br, and Cl) by stirring for 30 min a solution mixture that contained $Bi(NO_3)_3{\cdot}5H_2O$ dissolved in EG and inorganic salts of the respective halogens (KI, NaBr, and KCl). Subsequently, the precipitates were collected, subjected to a washing process, and afterward dried at a temperature of 50 °C [11]. Additionally, the growth orientation of BiOX crystals is influenced by the differences in solvents that occur along this pathway. Xiong and colleagues investigated the effect of alcohol chain length on the thermal production of BiOBr [12]. They did this by heating n-hexanol, n-butanol, and methanol as solvents for 5 h at a temperature of 180 °C in a tank made of stainless steel. The enhancement of the carbon chain length in n-alcohol resulted in a more pronounced and resilient manifestation of the characteristic diffraction patterns of BiOBr at the (110) as well as (102) facets. This observation suggests an improvement in the crystallization process of BiOBr.

According to previous research, it has been determined that BiOBr, specifically with its (001) facet exposed, exhibits enhanced PCA. The reason for this occurrence can be ascribed to the observation that the (001) facet exhibits a surface energy that is notably lower compared to the other facets. Additionally, BiOBr materials possess

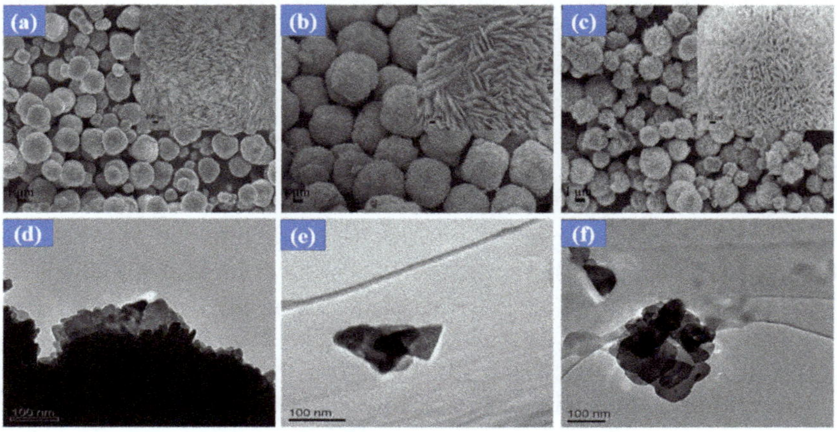

Figure 2.1. SEM image of (a) BiOCl, (b) BiOBr, and (c) BiOI and TEM images of (d) BiOCl, (e) BiOBr, and (f) BiOI. (Reproduced with permission from [11]. Copyright 2008 American Chemical Society.)

an intraelectric field oriented along the [001] direction [13]. The utilization of an ionic liquid, specifically Br, as opposed to KBr, was the determining factor in the synthesis of BiOBr with an exposed (001) facet, owing to the passivating properties of the ionic liquid and its ability to decrease the surface energy of the facet [14].

Surfactants like hexadecyl-trimethyl-ammonium bromide and polyvinylpyrrolidone (PVP) can purportedly modulate morphology when added to a substance. Utilizing the solvothermal method, Liu *et al* were able to create BiOBr floral nanospheres with a surface area of 32.9 m^2 g^{-1} by utilizing $Bi(NO_3)_3{\cdot}5H_2O$, EG, bromide [C16min]Br, and PVP ($M = 45\,000$–$58\,000$) as precursors and maintaining the reaction at a temperature of 140 °C for 24 h [15]. Hao *et al* investigated the influence of PVP on the PCA of BiOI [16]. The BiOI NSs exhibited self-aggregation into 3D hierarchical MSs in the presence of PVP. These MSs demonstrated a better PCA in destroying TC-HC than the 2D-BiOI NSs (a removal rate of 94% versus 44%). Mera *et al* [17] successfully prepared two distinct types of BiOI by employing 1-butyl-3-methylimidazolium iodide ([bmim]I) and KI as sources of iodine in the appropriate proportions. The BiOI synthesized using [bmim]I exhibited the highest PCA results for removing gallic acid, which could be assigned to the significantly exposed (001) crystal facets. Jiang *et al* successfully manufactured BiOI similar to a hollow flower morphology (h-BiOI) with the help of EG solvent [18]. A 2 nm thick layer of h-BiOI NSs showed an improved oxidation capacity and larger specific surface area compared to bulk BiOI. As a result, they were able to degrade 99% of RhB within 60 min, which is significantly quicker than the rate at which bulk BiOI could do so. Furthermore, Zhang *et al* [19] created nanoplates of $BiOBr_xI_{1-x}$ with the (001) crystal plane being exposed. The band gap was lowered from 2.87 to 1.89 eV by lowering the ratio of Br/I, which increased the amount of light that was absorbed. The optimized $BiOBr_{0.8}I_{0.2}$ displayed the maximum PCA for the decolorization of RhB. This activity was 1.7 and 5.4 times greater than BiOBr and BiOI, respectively.

The kind of solvent that is used in the solvothermal process influences the performance, surface structure, and morphology of BiOCl. Using EG as a solvent, Zhao *et al* [20] produced BiOCl containing oxygen vacancies (OV-BOC). Compared to pure BiOCl which was synthesized using ethanol, the OV-BOC exhibited a better capacity for oxidation, which resulted in an efficient photocatalytic breakdown of phenol, RhB, and MO. In a separate piece of research, the synthesis of BiOCl photocatalysts with various morphologies was carried out using ethanol, glycerol, and EG as the solvents of choice [21]. Based on the comparison results, hierarchical MSs of BiOCl produced in EG solvent demonstrated the best PCA for the breakdown of carbamazepine (CBZ). This was primarily attributable to their high separation efficiency and higher adsorption capability. Xing *et al* [22] used water, ethanol, isobutanol, EG, and glycerin as the starting materials for synthesizing a variety of BiOBr photocatalysts. The BiOBr having flower-like morphology synthesized in glycerin exhibited the highest PCA for the breakdown of reactive brilliant blue (KN-R). This might be credited to its abundant oxygen vacancies, exposed active surface, and small particle size. Using ethanol, EG, 2-methoxyethanol, benzyl alcohol, and tert-butanol as the solvents, Wei *et al* [23] successfully produced some different BiOBr photocatalysts. As a consequence of this, BiOBr thin NSs produced in benzyl alcohol displayed a remarkable PCA for the destruction of RhB, which has been ascribed to the exposed (001) facets on the surface of the NSs. In addition, a variety of BiOI photocatalysts were produced by employing water, ethanol, EG, and glycerin, respectively, in the synthesis process [24]. The three-dimensional mesoporous hierarchitectured BiOI that was produced in glycerol showed exceptional PCA for the elimination of As(III) under sunlight. This was owing to the sample's large specific surface area, highly exposed (001) facets, and mesoporous structure. Dehghan *et al* [25] used the hydrolysis procedure and the solvothermal method to produce two types of BiOI photocatalysts [25]. Regarding the breakdown of TC-HCl, the sample generated utilizing the solvothermal approach revealed a clear advantage. By incorporating PEG1000 into PEG400 solvent, Guo *et al* [26] created a 3D mesoporous nanostructured BiOCl. The PCA for the elimination of bisphenol A (BPA) was significantly increased due to abundant vacancies and mesopores produced on thin BiOCl NSs (3–6 nm in thickness).

The ratio of water to methanol, as well as the type of alcohol employed, influence both the performance and morphology of BiOCl. By changing the proportions of water to methanol, for instance, Xu *et al* [27] produced BiOCl nanosheets of varying dimensions and forms. When the water concentration was at 10%, the BiOCl that was synthesized exhibited a greater PCA in the decolorization of MO as well as RhB. Furthermore, Lee *et al* [28] utilized an ethanol/EG combination as the solvent for the synthesis of BiOX MSs. When flower-like BiOI MSs were subjected to photocatalytic hydrogen evolution, they demonstrated the maximum PCA towards the lowest fluorescence intensity.

2.3 Hydrolysis method

The production of BiOX, which depends on the reaction between a Bi salt (e.g. BiX_3 and $Bi(NO_3)_3$) with water or an oxyhalide [29], may also be accomplished by hydrolysis, which is a popular approach. One notable benefit of employing this particular approach is its ability to facilitate the straightforward synthesis of BiOX particles of diverse sizes within a conventional reactor set-up. In contrast to the hydro/solvothermal approach, the synthesis process that uses hydrolysis allows for the use of a wider variety of bismuth salts as precursors. These salts include halides, nitrates, and oxides, among others. However, even when the circumstances of the reaction are not very harsh, it is difficult to maintain control of the reaction to produce products with dimensions that are consistent across the board [30]. The precursor chemicals have a considerable impact on the shape as well as the yield of the catalysts during their production. In addition, the solvent affects the development of the crystals, which is an essential component of the photocatalytic performance [31]. In the study by Su *et al*, bismuth triiodide (BiI_3) was employed as a precursor in the preparation of BiOI, which resulted in the formation of a hierarchical flower-like structure [32]. The use of direct hydrolysis accomplished this synthesis. The utilization of hydrolysis as a method for fabricating photoelectrodes through the application of electrodeposition techniques is deemed suitable. Furthermore, the reaction media employed in this process can also serve as an electrochemical medium for modifying the electrode. In addition, Armelao *et al* created nanostructures of BiOCl by hydrolyzing $BiCl_3$ for 6 h at a temperature of 65 °C. Then, under acidic circumstances, acetylacetone was used as an auxiliary solvent to stabilize these nanostructures [33]. Similarly, Song *et al* found that thick NSs (21–85 nm) of BiOCl could be generated by hydrolyzing $Bi(NO_3)_3$ with Na_2CO_3 and HCl while keeping a pH of less than 2 for 30 min at ambient temperature [34]. At room temperature, Zhang *et al* introduced 0.5 g of $BiBr_3$ to 15 ml of a solvent of either isopropyl alcohol or water [35]. The researchers then analyzed the PCA of BiOBr that had been generated via hydrolysis and alcoholysis. The suspension was then heated for 10 minutes at temperatures of 20 °C, 40 °C, or 60 °C, while the pH was maintained at 9. The BiOBr that was produced by hydrolysis took the form of a nanosheet and possessed the highest possible crystallinity as well as (102) distinct facets. In contrast, the BiOBr produced through alcoholysis took the form of a flower and has (110) exposed facets. Both demonstrated a better PCA toward the breakdown of MO, while the BiOBr that resulted from hydrolysis showed greater stability after four cycles. Through hydrolysis performed at ambient temperature, Huang *et al* [36] were able to create BiOBr in a variety of configurations, including spherical quantum dots of zero dimension, nanorods of one dimension, nanosheets of two dimensions, and hierarchical structures of three dimensions (figure 2.2(a)). During the hydrolysis

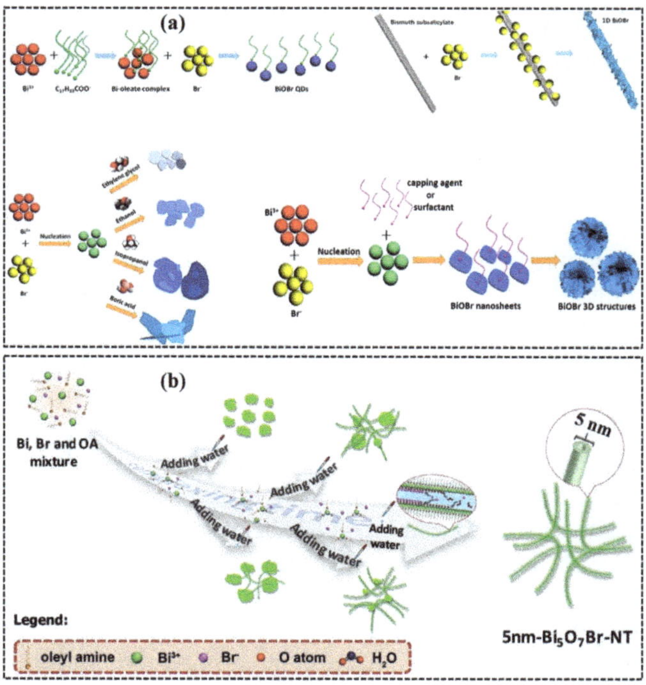

Figure 2.2. (a) Formation process of 0D BiOBr quantum dots, 2D BiOBr, and 3D BiOBr. (Reproduced with permission from [36]. Copyright 2020 Elsevier.) (b) Preparation process of ultrafine Bi_5O_7Br nanotubes. (Reproduced with permission from [37]. Copyright 2017 John Wiley and Sons.)

process, the morphology of BiOBr was altered by a variety of factors, including capping agents, different sources of bromine or bismuth, surfactants, and solvents. For the manufacture of ultrafine Bi_5O_7Br nanotubes with a 5 nm diameter, Wang *et al* [37] utilized a hydrolysis process, which consisted of two primary steps: (i) by combining bismuth ions and bromine ions in oleylamine a complex known as bismuth–bromo–oleylamine (Bi–Br–OA) was produced and (ii) by gradually adding water to the complex it was hydrolyzed, generating ultrafine nanotubes (figure 2.2(b)). Surfactants are typically added to catalysts synthesized via the hydrolysis process since they generally have low dispersibility. Consequently, the hydrolysis process depends on the solvent.

2.4 Precipitation method

The chemical precipitation approach has been utilized extensively in the production of BiOX photocatalysts as a result of its straightforward synthesis conditions, high efficiency, and low amount of required energy [38, 39]. In one study, an alkali solution used in the conventional precipitation approach was progressively poured into the Bi-containing precursors as the process was being carried out. To effectively manufacture a 3D hierarchical structure made of BiOI nanoplates using chemical precipitation at room temperature, Li *et al* employed KI as the halogen source and $Bi(NO_3)_3{\cdot}5H_2O$ as the bismuth source [40]. Under VL, the BiOI that was produced exhibited efficient PCA for the destruction of model pollutants, including MO and phenol. Although it appears to have many benefits, chemical precipitation also has many drawbacks. These drawbacks include impurity generation, uncontrolled morphology, particle agglomeration, and so on. In an acetic acid water solution, KBr and $Bi(NO_3)_3$ were stirred and mixed for 6 h to synthesize BiOBr using the method of co-precipitation [41]. A simple co-precipitation method has been successfully employed to create BiOCl microflowers, and it was discovered that their PCA performance in acidic environments was superior to that in neutral and alkaline environments [42]. Additionally, BiOI was produced via co-precipitation, and EDTA was utilized both as a structure-directing agent and as a reaction retarder in aqueous media. The BiOI sample obtained in this study had a significant BET surface area of around 47.5 $m^2\ g^{-1}$. Consequently, it demonstrated exceptional PCA for removing NO in the gaseous phase, achieving an impressive efficiency of nearly 98% [43]. Furthermore, Mustajab *et al* produced pure BiOI and PVP-doped BiOI quantum dots by employing the co-precipitation approach. They utilized potassium iodide (KI) and $Bi(NO_3)_3{\cdot}5H_2O$ as their sources of bismuth and iodine. According to the findings of the study, the PVP-doped version of BiOI showed superior effectiveness in the removal of methylene blue (MB) dye compared to the unmodified version. PVP doping resulted in a reduction in the size of the quantum dots, which led to a greater surface area and thus an increase in MB degradation rate [44]. The preparation of BiOCl, BiOI, and BiOBr photocatalysts was carried out by Zhang *et al* [45], following this approach. The activity sequence observed during the elimination of Hg^0 under the influence of fluorescent light shows that the photocatalytic performance of BiOI is superior in comparison to BiOCl and BiOBr, with BiOI demonstrating the highest efficacy. Huang *et al* [48] effectively synthesized the photocatalyst Br–BiOI by incorporating Br- into the BiOI lattice [46]. Incorporating Br into BiOI led to a reduction in the potential of the valence band along with an enlargement of the band gap. This prevented the photogenerated charge carriers from recombining and led to an increase in the PCA of BiOI. In addition, by adjusting the band engineering, the PCA of BiOX may be made more effective. In this regard, powders of $Bi_{24}O_{31}Cl_{10}$ were produced using chemical precipitation [47]. Because $Bi_{24}O_{31}Cl_{10}$ has a narrow band gap, it exhibited enhanced PCA for the decolorization of RhB when exposed to VL. In addition, BiOI samples were manufactured using the co-precipitation technique in a glycerol medium at a temperature of 100 °C. Subsequently, the precursor powders underwent thermal treatment at a temperature of 250 °C [48]. Additionally, the polyols sorbitol and mannitol were added to modify the influence that –OH groups had on the final product's physical and chemical properties. Following a series of characterization experiments, the researchers observed that the introduction of mannitol led to the development of BiOI NSs that displayed distinctive anisotropic growth along the crystal orientation (102). Figure 2.3 illustrates a potential growth process that was hypothesized based on the findings of experiments. When the compound $Bi(NO_3)_3{\cdot}5H_2O$ is dissolved in glycerol, a complex of bismuth called Bi-$(C_3H_8O_3)$ is produced. Once the reaction attains a suitably elevated temperature, the Bi complex has the potential to undergo dissolution, leading to the liberation of BiO^+ ions into the surrounding solution. Simultaneously, the generation of I ions occurs due to the dissolution of KI prompted by the presence of glycerol inside the medium. Consequently, the ions of BiO^+ that are released into the solution react with the pre-existing ions of I^-, leading to the creation of BiOI nuclei. The color of the solution undergoes a gradual transition from transparent to yellow due to the progressive accumulation

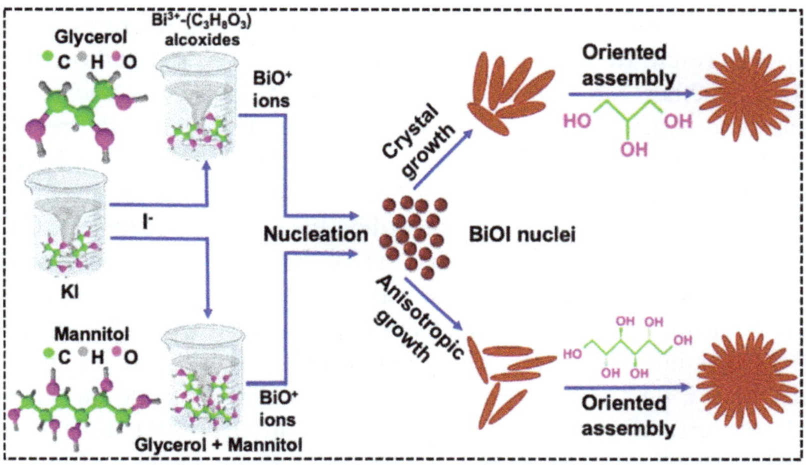

Figure 2.3. The possible growth mechanism of BiOI hierarchical structures. (Reproduced with permission from [48]. Copyright 2022 Springer.)

of BiOI crystal nuclei in the suspension throughout the course of the reaction. Because of the inherently crystalline character of BiOX, the crystal tends to develop in the form of NSs. These NSs often self-assemble into hierarchical patterns to reduce their surface energy. In addition to acting as a solvent, glycerol may also acts as a chelating agent, facilitating the formation of BiO^+ ions by aiding in the development of the bismuth complex. Similar to this, it performs the role of a soft template by directing the formation of aggregated NSs and their directed assembly to produce BiOI hierarchical structures [48].

2.5 Ultrasonic/microwave-assisted method

For the synthesis of a wide variety of nanomaterials in a range of controlled sizes and shapes, a process assisted by microwaves is considered to be an approach that is practical, fast, gentle, and effective [49]. This approach utilizes electromagnetic waves within the wavelength range from 1 mm to 1 m, which falls between the infrared as well as radio regions in the electromagnetic radiation spectrum. Heat is transferred from the microwave process to the material by ion conduction and dipole rotation, respectively. Moreover, the procedure exhibits a substantial yield, and the limited production of by-products can be regulated and harnessed to facilitate specific reactions and standards with greater ease. This is something that is not possible with the more typical heating methods.

With the aid of the microwave method, Ai and coworkers developed hierarchical porous BiOI. Under the influence of light irradiation, the as-obtained BiOI is easily renewable and has a high capacity for adsorption towards Congo red in water [50]. Zhou *et al* [51] used a straightforward microwave technique to manufacture n–n heterojunction composites of $La(OH)_3$/BiOCl (BCL). The findings demonstrated that, within a specific range, the thickness of the $La(OH)_3$/BCL NSs decreased as the concentration of La source increased. This suggested that $La(OH)_3$ can inhibit BiOCl growth. It is commonly understood that shorter migration paths caused by thinner NSs would lower photogenerated carriers recombination rates, boosting the PCA [52]. As a result, BCL-20 had the highest level of PCA. In addition, Li *et al* [53] investigated whether the size and shape of BiOX can be modified easily by modifying the quantity of halide, the concentration of the reaction precursor, and the amount of mannitol in the mixture. Images obtained using the SEM and TEM techniques demonstrated that increased halide concentration resulted in a larger average size of crystal for the BiOX. Regarding the reactant precursors, CTAB and CTAC served dual roles in the production of BiOX. They acted as sources of halogen for the synthesis of BiOX and also functioned as soft templates. This facilitated the formation of flower-like hierarchical nanostructures by promoting the assembly of NSs into a hierarchical arrangement through the interaction between the petals and CTA^+ ions. In addition, when the concentration of mannitol increased, the crystal's anisotropic development in the (001) direction became constricted gradually. This resulted in the (001) plane having a relatively low intensity. In addition, Maisang *et al* [54] examined the influence of varying the quantity of PVP utilized in the production of flower-like BiOBr/BiOCl composite. The size of the BiOBr/BiOCl composite increased with the addition of PVP, which ranged from 0 to 1.50 g. Additionally, the particles began to cluster together. When there was a concentration of PVP equal to one gram, the sample took on the appearance of flower-like particles with a size of less than 1500 nm and exhibited exceptional PCA.

Intaphong *et al* employed a conventional chemical bath method to synthesize nanoplates of BiOI through a sonochemical process conducted in an ultrasonic bath operating at a temperature of 80 °C. The nanoplates of BiOI prepared under alkaline conditions with a pH = 12 exhibited the most significant level of efficacy in destroying RhB molecules when exposed to VL [55].

2.6 Two-phase method

In addition to more conventional approaches to synthesis, a two-phase reaction method has been utilized to produce BiOCl NSs at the water–air interface [56]. The NSs of BiOCl that resulted, with their exposed (010) facet, exhibited a strong PCA for the breakdown of MO. In yet another two-phase technique [57], octadecene (ODE), oleic acid (OA), and oleylamine (OLA) were used to dissolve Bi $(NO_3)_3{\cdot}5H_2O$ in the oil phase, while an aqueous solution of KX was used in the water phase. Ultrathin two-dimensional BiOX NSs were synthesized by refluxing a mixed solution at a temperature of 170 °C. The preparation technique facilitated the creation of an ultrathin BiOX structure characterized by a significant exposure of {001} facets, owing to the continuous presence of an acidic environment.

2.7 Physical method

For the synthesis of BiOX photocatalyst, it has been reported that several physical processes such as ultrasonic exfoliation, mechanical grinding, and calcination or combustion can be used.

2.7.1 Combustion/calcination

When a BiOX phase transition to a BiOX rich in Bi is desired, calcination is a useful method. This is because heating causes the BiOX's crystal structure to become free of unstable halogen atoms during the calcination process. In the crystal structure of BiOX, there is a van der Waals interaction, although a weak one, between the two layers of halogen atoms and the layer of $[Bi_2O_2]^{2+}$. Because of this, calcination is required to eliminate some of the unstable halogen atoms, which ultimately results in a phase change from BiOX to Bi-rich BiOX. Two distinct types of calcination can be distinguished: (i) the full calcination reaction of pure BiOX, which would result in the end products with various bismuth contents, $Bi_xO_yX_z$, and (ii) the partial calcination reaction of pure BiOX, which would result in the BiOX heterojunction photocatalyst. For example, by heating BiOBr flakes to 750 °C, Yu *et al* developed plate-like $Bi_{24}O_{31}Br_{11}$ and tested its PCA for the degradation of Acid Orange II [58]. In addition, several other morphologies have been seen, as opposed to the plate-like appearance that was previously reported. Hollow $Bi_{24}O_{31}Cl_{10}$ microspheres were produced by subjecting BiOCl to a thermal treatment at a temperature of 600 °C in an environment of carbon-based materials which functioned as sacrificial template [59]. Recently, to produce a three-dimensional flower-sphere $BiOBr/Bi_4O_5Br_2$-OV heterojunction, Li *et al* used a NaOH-modified three-dimensional flower-like BiOBr precursor [60]. They studied its PCA for TC and CIP degradation. The boost of photocatalytic performance was achieved through significant improvements in the separation of photogenerated carriers of charge, facilitated by constructing well-alloyed surfaces (figure 2.4(a)). Using this procedure, Huang *et al* [61] successfully

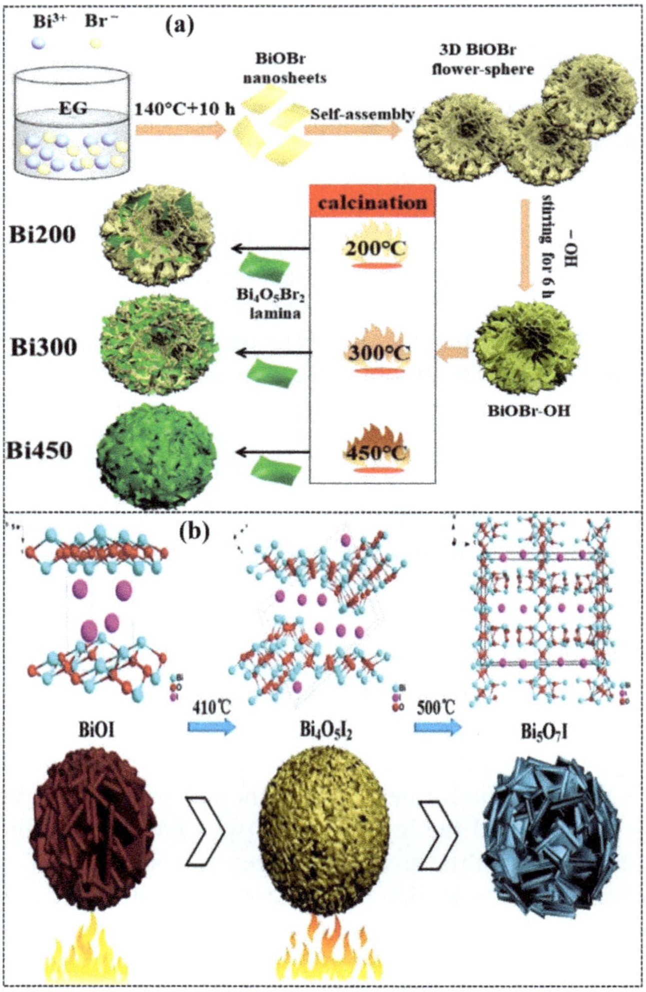

Figure 2.4. (a) Synthesis of the 3D flower-sphere structure of BiOBr, Bi200, Bi300, and Bi450. (Reproduced with permission from [60]. Copyright 2020 Elsevier.) (b) Schematic illustration of the hierarchical architectures of BiOI, $Bi_4O_5I_2$, $Bi_4O_5I_2$–Bi_5O_7I composite, and Bi_5O_7I. (Reproduced with permission from [61]. Copyright 2017 Elsevier.)

synthesized $Bi_4O_5I_2$, Bi_5O_7I, and $Bi_4O_5I_2$–Bi_5O_7I composites by starting with BiOI as the precursor (figure 2.4(b)). Temperatures of 350 °C, 380 °C, 410 °C, 440 °C, 470 °C, and 500 °C were used for the calcination process. After heating, the specimens were gathered and analyzed using XRD techniques. The resultant diffraction patterns provided insights into the phase transition of BiOI. In a different piece of research, Lee *et al* [62] reported the preparation of Bi-rich composites based on BiOI using hydrothermal methods without the requirement of calcination.

2.7.2 Ultrasonic exfoliation method

The fabrication of monolayered BiOBr NSs, with 0.85 nm thickness, was achieved through an ultrasonic exfoliation approach [63]. The NSs were fabricated in formamide solvent. The performance of the monolayered BiOBr was superior in terms of both adsorption and photodegradation. It was possible to construct hierarchical flower-like nanostructures of BiOCl, BiOBr, and BiOI with a thickness of 5 min using a straightforward mechanical grinding approach that did not need any solvents [64]. Three different samples of BiOX exhibited remarkable PCA, particularly in the breakdown of MB and RhB. The approach of grinding without the use of solvents may be implemented to obtain mass production of BiOX. This is because the preparation time is very fast, and the PCA is quite high.

2.7.3 Template method

The principal mechanism by which this approach alters the morphology of BiOX is the modification of the growth process and crystal nucleation that takes place throughout the synthesis. When using the template approach, the synthesis of nanomaterials typically consists of three stages [65]. First, a template is created. Second, the target product is created by synthesizing it under the influence of the template by using a standard synthesis process such as precipitation, sol–gel, hydrothermal, or another approach. Third, the template is removed. Carbonaceous microspheres were used by Cui *et al* as a template to adsorb bivalent cations (Bi^{3+}) and anionic cations (Cl). After adsorption and sintering at temperatures reaching 400 °C, the researchers obtained hollow BiOCl MSs with a diameter of 200 nm and a shell thickness of 40 nm [59]. Yan *et al* [66] employed a biological template, specifically a butterfly wing, to synthesize hierarchical BiOCl. In addition, BiOI was utilized as a self-sacrifice template to create $Bi_4O_5I_2/Bi_5O_7I$ and $BiOI/Bi_4O_5I_2$ heterojunctions by the process of simple calcination [67]. To create $BiOBr/Bi_4O_5Br_2$, the BiOBr self-sacrifice template was also utilized [60]. Table 2.1 presents recent studies conducted on the synthesis of BiOX materials.

Table 2.1. Summary of typical synthesis methods of BiOX in different morphologies.

Processing	Raw material and synthesis conditions	Product	Morphology	References
Mechanical grinding	KX, and $Bi(NO_3)_3 \cdot 5H_2O$, dried at 60 °C	BiOX (X = Cl, Br, I)	Flower-like hierarchical microsphere/ nanostructures	[64]
Chemical precipitation	EG, KI, $Bi(NO_3)_3 \cdot 5H_2O$, water, dried 60 °C for 4 h	BiOI	Spherical microstructure self-assembled by abundant ultrathin nanosheets	[68]
Hydrothermal method	$Bi(NO_3)_3 \cdot 5H_2O$, PVP, EG, KI, NaBr, autoclaved at 160 °C for 12 h and dried at 70 °C for 6 h	BiOI/BiOBr composite	3D hierarchical microsphere	[69]
Hydrothermal method	$Bi(NO_3)_3 \cdot 5H_2O$, $C_2H_6BrN \cdot HBr$, $C_{10}H_5NbO_{20} \cdot xH_2O$	Nb–BiOBr	Microsphere	[70]
Chemical transformation technique	Bi_2O_3, HI, dried at 80 °C for 12 h	BiOI	Nanosheets	[71]
Hydrothermal method	$Bi(NO_3)_3 \cdot 5H_2O$, HNO_3, glucose, KI, 24 h of autoclaving at 160 °C and 18 h of drying at 60 °C	Carbon–BiOI	Blooming flower which is composed of many nanosheets	[72]
Hydrothermal method	$Bi(NO_3)_3 \cdot 5H_2O$, $In(NO_3)_3$, KI, 2 h of autoclaving at 160 °C and 12 h of drying at 60 °C	In–BiOI	Nanosheets	[73]
Hydrothermal method	$Bi(NO_3)_3 \cdot 5H_2O$, HNO_3, NaXs (X = Cl, Br, I), dried at 50 C	BiOX	Smooth lamellar structure	[74]
Hydrothermal method	$Bi(NO_3)_3 \cdot 5H_2O$,PVP, NaCl, EG, 8 h of autoclaving at 160 °C and drying in air at 60 °C	BiOCl	Nanosheets	[75]
Solvothermal route	$BiCl_3$, pyridine, dried for 6 h in a vacuum at 60 °C	BiOCl	Flower-like	[76]
Hydrothermal process	NaCl, $Bi(NO_3)_3 \cdot 5H_2O$, mannitol, EG, diethylene glycol, 3 h of autoclaving at 150 °C	BiOCl	Square-like nanoplates	[77]
Solvothermal method	CTAB, $Bi(NO_3)_3 \cdot 5H_2O$, EG, 12 h of autoclaving at 180 °C and 6 h of drying at 60 °C	BiOBr	Microspheres with rough surface	[78]
Solvothermal method	$Bi(NO_3)_3 \cdot 5H_2O$, ethanol, CTAB, autoclaved at 150 °C for 24 h and dried in air	BiOBr	Flower-like microspheres, which are composed of many radially grown nanosheets	[79]
Hydrolysis method	Bi_2O_3, halogen acid, heated at 100 °C	BiOX	Homogeneous sheet-shaped BiOX crystals	[80]

(Continued)

Table 2.1. (*Continued*)

Processing	Raw material and synthesis conditions	Product	Morphology	References
Solvothermal process	$Bi(NO_3)_3·5H_2O$, NaCl, autoclaved for 6 h at 140 °C	BiOCl	Microspheres	[81]
Solvothermal	$Bi(NO_3)_3·5H_2O$, KCl, EG, autoclaved at 160 °C for 12 h and dried in an air oven at 50 °C	BiOCl	Hierarchical microspheres composed by polymerizated BiOCl nanosheet	[82]
Precipitation	$NaBiO_3$ and HX	BiOX	Sheet-like	[83]
Reverse microemulsions	$Bi(NO_3)_3·5H_2O$, NaCl, KBr, or KI	BiOX	Nanoparticles	[84]
Molecular precursor route	$BiCl_3$, thiourea (Tu), 80 °C for 12 h	BiOCl	Nanosheets	[85]
Mini-emulsion-mediated route	2-methoxyethanol, ([C16Mim]Br), Bi $(NO_3)_3·5H_2O$, overnight dried at 60 °C in vacuum	BiOBr	Quasi-uniform hierarchical microspheres	[86]
Solvothermal	$Bi(NO_3)_3·5H_2O$, CTAB, DEG, dried at 60 °C in vacuum for 6 h	BiOBr	Microspheres	[87]
Solvothermal	[Bmim]I, $Bi(NO_3)_3·5H_2O$; 140 °C for 24 h	BiOI	Sphere-like	[88]
Low-temperature water-assisted self-assembly	KBr, $Bi(NO_3)_3·5H_2O$, oleamine	Bi_5O_7Br	Nanotubes	[37]
Water-induced self-assembly	$Bi(NO_3)_3·5H_2O$, ammonia, KBr, 20 °C, 40 °C, 60 °C, 80 °C, and 100 °C	Bi_5O_7Br	Nanotubular structures	[89]
Annealing	BiI_3, 350 °C for 3 h and 280 °C for 16 h	BiOI	Single-crystal nanosheets	[90]
Annealing	$BiCl_3$, annealed at 400 °C for 2 h	BiOCl	Nanosheets	[91]
Molecular precursor and hydrolytic process	KI, $Bi(NO_3)_3·5H_2O$, glycerol, dried in air at 80 ° C and heated at 300 C for 5 h	$Bi_4O_5I_2$	Particles	[92]
Molecular precursor and calcination	KI, $Bi(NO_3)_3·5H_2O$, glycerol, dried in air at 80 ° C and calcinated at 450 C for 5 h	Bi_5O_7I	Particles	[92]
Molecular precursor hydrolysis or calcination	KI, $Bi(NO_3)_3·5H_2O$, glycerol, calcinated at 400 °C	Bi_5O_7I	Nanosheets	[93]
Co-precipitation	$Bi(NO_3)_3·5H_2O$, KI, dried at for 12 h at 150 °C	BiOI	Quantum dots	[44]

References

[1] Wu X, Toe C Y, Su C, Ng Y H, Amal R and Scott J 2020 Preparation of Bi-based photocatalysts in the form of powdered particles and thin films: a review *J. Mater. Chem.* A **8** 15302–18

[2] Jiang Z, Yang F, Yang G, Kong L, Jones M O, Xiao T and Edwards P P 2010 The hydrothermal synthesis of BiOBr flakes for visible-light-responsive photocatalytic degradation of methyl orange *J. Photochem. Photobiol.* A **212** 8–13

[3] Zhang J *et al* 2017 The effect of pH on synthesis of BiOCl and its photocatalytic oxidization performance *Mater. Lett.* **186** 353–6

[4] Zhang X X, Li R, Jia M, Wang S, Huang Y and Chen C 2015 Degradation of ciprofloxacin in aqueous bismuth oxybromide (BiOBr) suspensions under visible light irradiation: a direct hole oxidation pathway *Chem. Eng. J.* **274** 290–7

[5] Intaphong P, Phuruangrat A, Thongtem T and Thongtem S 2020 Effect of pH on phase, morphologies, and photocatalytic properties of BiOCl synthesized by hydrothermal method *J. Aust. Ceram. Soc.* **56** 41–8

[6] Feng H *et al* 2015 Modulation of photocatalytic properties by strain in 2D BiOBr nanosheets *ACS Appl. Mater. Interfaces* **7** 27592–6

[7] Dou L, Ma D, Chen J, Li J and Zhong J 2019 F127-assisted hydrothermal preparation of BiOI with enhanced sunlight-driven photocatalytic activity originated from the effective separation of photo-induced carriers *Solid State Sci.* **90** 1–8

[8] Zou Z, Xu H, Li D, Sun J and Xia D 2019 Facile preparation and photocatalytic activity of oxygen vacancy rich BiOCl with {0 0 1} exposed reactive facets *Appl. Surf. Sci.* **463** 1011–8

[9] Li K, Liang Y, Yang J, Gao Q, Zhu Y, Liu S, Xu R and Wu X 2017 Controllable synthesis of {001} facet dependent foursquare BiOCl nanosheets: a high efficiency photocatalyst for degradation of methyl orange *J. Alloys Compd.* **695** 238–49

[10] Cai Y, Li D, Sun J, Chen M, Li Y, Zou Z, Zhang H, Xu H and Xia D 2018 Synthesis of BiOCl nanosheets with oxygen vacancies for the improved photocatalytic properties *Appl. Surf. Sci.* **439** 697–704

[11] Zhang X, Ai Z, Jia F and Zhang L 2008 Generalized one-pot synthesis, characterization, and photocatalytic activity of hierarchical BiOX (X = Cl, Br, I) nanoplate microspheres *J. Phys. Chem.* C **112** 747–53

[12] Xiong X, Ding L, Wang Q, Li Y, Jiang Q and Hu J 2016 Synthesis and photocatalytic activity of BiOBr nanosheets with tunable exposed (0 1 0) facets *Appl. Catal.* B **188** 283–91

[13] Zhao L, Zhang X, Fan C, Liang Z and Han P 2012 First-principles study on the structural, electronic and optical properties of BiOX (X = Cl, Br, I) crystals *Physica* B **407** 3364–70

[14] Mao D, Lü X, Jiang Z, Xie J, Lu X, Wei W and Showkot Hossain A M 2014 Ionic liquid-assisted hydrothermal synthesis of square BiOBr nanoplates with highly efficient photocatalytic activity *Mater. Lett.* **118** 154–7

[15] Liu Y and Wu Q 2017 One novel material with high visible-light activity: hexagonal Cu flakelets embedded in the petals of BiOBr flower-nanospheres *J. Nanopart. Res.* **19** 1–14

[16] Hao R, Xiao X, Zuo X, Nan J and Zhang W 2012 Efficient adsorption and visible-light photocatalytic degradation of tetracycline hydrochloride using mesoporous BiOI microspheres *J. Hazard. Mater.* **209–10** 137–45

[17] Mera A C, Moreno Y, Contreras D, Escalona N, Meléndrez M F, Mangalaraja R V and Mansilla H D 2017 Improvement of the BiOI photocatalytic activity optimizing the solvothermal synthesis *Solid State Sci.* **63** 84–92

[18] Jiang Z, Liang X, Liu Y, Jing T, Wang Z, Zhang X, Qin X, Dai Y and Huang B 2017 Enhancing visible light photocatalytic degradation performance and bactericidal activity of BiOI via ultrathin-layer structure *Appl. Catal.* B **211** 252–7
[19] Zhang X, Wang C Y, Wang L W, Huang G X, Wang W K and Yu H Q 2016 Fabrication of $BiOBr_xI_{1-x}$ photocatalysts with tunable visible light catalytic activity by modulating band structures *Sci. Rep.* **6** 22800
[20] Zhao H, Liu X, Dong Y, Xia Y and Wang H 2019 A special synthesis of BiOCl photocatalyst for efficient pollutants removal: new insight into the band structure regulation and molecular oxygen activation *Appl. Catal.* B **256** 117872
[21] Gao X, Guo Q, Tang G, Zhu W and Luo Y 2019 Controllable synthesis of solar-light-driven BiOCl nanostructures for highly efficient photocatalytic degradation of carbamazepine *J. Solid State Chem.* **277** 133–8
[22] Xing H, Ma H, Fu Y, Zhang X, Dong X and Zhang X 2015 Preparation of BiOBr by solvothermal routes with different solvents and their photocatalytic activity *J. Renew. Sustain. Energy* **7** 153103
[23] Wei X X, Cui B, Wang X, Cao Y Z, Gao L B, Guo S and Chen C M 2019 Tuning the physico-chemical properties of BiOBr via solvent adjustment: towards an efficient photocatalyst for water treatment *CrystEngComm.* **21** 1750–7
[24] Hu J, Weng S, Zheng Z, Pei Z, Huang M and Liu P 2014 Solvents mediated-synthesis of BiOI photocatalysts with tunable morphologies and their visible-light driven photocatalytic performances in removing of arsenic from water *J. Hazard. Mater.* **264** 293–302
[25] Dehghan A, Dehghani M H, Nabizadeh R, Ramezanian N, Alimohammadi M and Najafpoor A A 2018 Adsorption and visible-light photocatalytic degradation of tetracycline hydrochloride from aqueous solutions using 3D hierarchical mesoporous BiOI: synthesis and characterization, process optimization, adsorption and degradation modeling *Chem. Eng. Res. Des.* **129** 217–30
[26] Guo S Q, Zhu X H, Zhang H J, Gu B C, Chen W, Liu L and Alvarez P J J 2018 Improving photocatalytic water treatment through nanocrystal engineering: mesoporous nanosheet-assembled 3D BiOCl hierarchical nanostructures that induce unprecedented large vacancies *Environ. Sci. Technol.* **52** 6872–80
[27] Xu Y, Hu X, Zhu H and Zhang J 2016 Insights into BiOCl with tunable nanostructures and their photocatalytic and electrochemical activities *J. Mater. Sci.* **51** 4342–8
[28] Lee G J, Zheng Y C and Wu J J 2018 Fabrication of hierarchical bismuth oxyhalides (BiOX, X = Cl, Br, I) materials and application of photocatalytic hydrogen production from water splitting *Catal. Today* **307** 197–204
[29] Li P, Gao S, Liu Q, Ding P, Wu Y, Wang C, Yu S, Liu W, Wang Q and Chen S 2021 Recent progress of the design and engineering of bismuth oxyhalides for photocatalytic nitrogen fixation *Adv. Energy Sustain. Res.* **2** 2000097
[30] Wei X, Akbar M U, Raza A and Li G 2021 A review on bismuth oxyhalide based materials for photocatalysis *Nanoscale Adv.* **3** 3353–72
[31] Li G, Wang X, Zhang L and Zhu C 2021 Electronic structures and optical properties of BiOBr/BiOI heterojunction with an oxygen vacancy *Chem. Phys.* **549** 111264
[32] Su J, Xiao Y and Ren M 2014 Direct hydrolysis synthesis of BiOI flowerlike hierarchical structures and its photocatalytic activity under simulated sunlight irradiation *Catal. Commun.* **45** 30–3

[33] Armelao L, Bottaro G, MacCato C and Tondello E 2012 Bismuth oxychloride nanoflakes: interplay between composition-structure and optical properties *Dalton Trans.* **41** 5480–5
[34] Song Z, Dong X, Wang N, Zhu L, Luo Z, Fang J and Xiong C 2017 Efficient photocatalytic defluorination of perfluorooctanoic acid over BiOCl nanosheets via a hole direct oxidation mechanism *Chem. Eng. J.* **317** 925–34
[35] Li R, Gao X, Fan C, Zhang X, Wang Y and Wang Y 2015 A facile approach for the tunable fabrication of BiOBr photocatalysts with high activity and stability *Appl. Surf. Sci.* **355** 1075–82
[36] Han L, Guo Y, Lin Z and Huang H 2020 0D to 3D controllable nanostructures of BiOBr via a facile and fast room-temperature strategy *Colloids Surf.* A **603** 125233
[37] Wang S, Hai X, Ding X, Chang K, Xiang Y, Meng X, Yang Z, Chen H and Ye J 2017 Light-switchable oxygen vacancies in ultrafine Bi_5O_7Br nanotubes for boosting solar-driven nitrogen fixation in pure water *Adv. Mater.* **29** 1701774
[38] Chang X, Gondal M A, Al-Saadi A A, Ali M A, Shen H, Zhou Q, Zhang J, Du M, Liu Y and Ji G 2012 Photodegradation of Rhodamine B over unexcited semiconductor compounds of BiOCl and BiOBr *J. Colloid Interface Sci.* **377** 291–8
[39] Ren K, Liu J, Liang J, Zhang K, Zheng X, Luo H, Huang Y, Liu P and Yu X 2013 Synthesis of the bismuth oxyhalide solid solutions with tunable band gap and photocatalytic activities *Dalt. Trans.* **42** 9706–12
[40] Li Y, Wang J, Yao H, Dang L and Li Z 2011 Efficient decomposition of organic compounds and reaction mechanism with BiOI photocatalyst under visible light irradiation *J. Mol Catal.* A **334** 116–22
[41] Lu L, Kong L, Jiang Z, Lai H H C, Xiao T and Edwards P P 2012 Visible-light-driven photodegradation of rhodamine B on Ag-modified BiOBr *Catal. Letters.* **142** 771–8
[42] Wang J and Zhang Z 2020 Co-precipitation synthesis and photocatalytic properties of BiOCl microflowers *Optik* **204** 164149
[43] Montoya-Zamora J M, Martínez-de la Cruz A and López Cuéllar E 2017 Enhanced photocatalytic activity of BiOI synthesized in presence of EDTA *J. Taiwan Inst. Chem. Eng.* **75** 307–16
[44] Mustajab M, Ikram M, Haider A, Ul-Hamid A, Nabgan W, Haider J, Ghaffar R, Shahzadi A, Ghaffar A and Saeed A 2022 Promising performance of polyvinylpyrrolidone-doped bismuth oxyiodide quantum dots for antibacterial and catalytic applications *Appl. Nanosci.* **12** 2621–33
[45] Zhang A, Xing W, Zhang D, Wang H, Chen G and Xiang J 2016 A novel low-cost method for Hg0 removal from flue gas by visible-light-driven BiOX (X = Cl, Br, I) photocatalysts *Catal. Commun.* **87** 57–61
[46] Huang H, Li X, Han X, Tian N, Zhang Y and Zhang T 2015 Moderate band-gap-broadening induced high separation of electron-hole pairs in Br substituted BiOI: a combined experimental and theoretical investigation *Phys. Chem. Chem. Phys.* **17** 3673–9
[47] Wang L *et al* 2014 A dye-sensitized visible light photocatalyst-$Bi_{24}O_{31}Cl_{10}$ *Sci. Rep.* **4** 1 8
[48] Reyna-Cavazos K A, de la Cruz A M, Contreras D and Longoria-Rodríguez F E 2022 Polyol-assisted coprecipitation synthesis of BiOI photocatalyst and its activity to remove NO_x *Res. Chem. Intermed.* **48** 949–67
[49] Yan Y, Sun S, Song Y, Yan X, Guan W, Liu X and Shi W 2013 Microwave-assisted *in situ* synthesis of reduced graphene oxide-$BiVO_4$ composite photocatalysts and their enhanced

photocatalytic performance for the degradation of ciprofloxacin *J. Hazard. Mater.* **250–1** 106–14
[50] Ai L, Zeng Y and Jiang J 2014 Hierarchical porous BiOI architectures: facile microwave nonaqueous synthesis, characterization and application in the removal of Congo red from aqueous solution *Chem. Eng. J.* **235** 331–9
[51] Zhou Z, Xu H, Li D, Zou Z and Xia D 2019 Microwave-assisted synthesis of $La(OH)_3$/BiOCl n–n heterojunctions with high oxygen vacancies and its enhanced photocatalytic properties *Chem. Phys. Lett.* **736** 136805
[52] Stoesser A, Von Seggern F, Purohit S, Nasr B, Kruk R, Dehm S, Wang D, Hahn H and Dasgupta S 2016 Facile fabrication of electrolyte-gated single-crystalline cuprous oxide nanowire field-effect transistors *Nanotechnology.* **27** 415205
[53] Li G, Qin F, Wang R, Xiao S, Sun H and Chen R 2013 BiOX (X = Cl, Br, I) nanostructures: mannitol-mediated microwave synthesis, visible light photocatalytic performance, and Cr (VI) removal capacity *J. Colloid Interface Sci.* **409** 43–51
[54] Maisang W, Promnopas S, Kaowphong S, Narksitipan S, Thongtem S, Wannapop S, Phuruangrat A and Thongtem T 2020 Microwave-assisted hydrothermal synthesis of BiOBr/BiOCl flowerlike composites used for photocatalysis *Res. Chem. Intermed.* **46** 2117–35
[55] Intaphong P, Phuruangrat A, Thongtem S and Thongtem T 2018 Sonochemical synthesis and characterization of BiOI nanoplates for using as visible-light-driven photocatalyst *Mater. Lett.* **213** 88–91
[56] Xu Z 2018 Synthesis of BiOCl nanosheets with exposed (010) facets via a facile two-phase reaction and photocatalytic activity *Ferroelectrics.* **527** 37–43
[57] Wang Z, Chu Z, Dong C, Wang Z, Yao S, Gao H, Liu Z, Liu Y, Yang B and Zhang H 2020 Ultrathin BiOX (X = Cl, Br, I) nanosheets with exposed {001} facets for photocatalysis *ACS Appl. Nano Mater.* **3** 1981–91
[58] Yu C, Zhou W, Yu J, Cao F and Li X 2012 Thermal stability, microstructure and photocatalytic activity of the bismuth oxybromide photocatalyst *Chin. J. Chem.* **30** 721–6
[59] Cui P, Wang J, Wang Z, Chen J, Xing X, Wang L and Yu R 2016 Bismuth oxychloride hollow microspheres with high visible light photocatalytic activity *Nano Res.* **9** 593–601
[60] Li P, Cao W, Zhu Y, Teng Q, Peng L, Jiang C, Feng C and Wang Y 2020 NaOH-induced formation of 3D flower-sphere $BiOBr/Bi_4O_5Br_2$ with proper-oxygen vacancies via *in situ* self-template phase transformation method for antibiotic photodegradation *Sci. Total Environ.* **715** 136809
[61] Huang H, Xiao K, Zhang T, Dong F and Zhang Y 2017 Rational design on 3D hierarchical bismuth oxyiodides via *in situ* self-template phase transformation and phase-junction construction for optimizing photocatalysis against diverse contaminants *Appl. Catal.* B **203** 879–88
[62] Lee W W, Lu C S, Chuang C W, Chen Y J, Fu J Y, Siao C W and Chen C C 2015 Synthesis of bismuth oxyiodides and their composites: characterization, photocatalytic activity, and degradation mechanisms *RSC Adv.* **5** 23450–63
[63] Yu H, Huang H, Xu K, Hao W, Guo Y, Wang S, Shen X, Pan S and Zhang Y 2017 Liquid-phase exfoliation into monolayered BiOBr nanosheets for photocatalytic oxidation and reduction *ACS Sustain. Chem. Eng.* **5** 10499–508
[64] Long Y, Han Q, Yang Z, Ai Y, Sun S, Wang Y, Liang Q and Ding M 2018 A novel solvent-free strategy for the synthesis of bismuth oxyhalides *J. Mater. Chem.* A **6** 13005–11

[65] Xie Y, Kocaefe D, Chen C and Kocaefe Y 2016 Review of research on template methods in preparation of nanomaterials *J. Nanomater.* **2016** 2302595

[66] Yan X *et al* 2019 *In situ* synthesis of BiOCl nanosheets on three-dimensional hierarchical structures for efficient photocatalysis under visible light *Nanoscale.* **11** 10203–8

[67] Cheng H, Huang Y, Wu J, Ling Y, Dong L, Zha J, Yu M and Zhu Z 2020 Controllable design of bismuth oxyiodides by *in situ* self-template phase transformation and heterostructure construction for photocatalytic removal of gas-phase mercury *Mater. Res. Bull.* **131** 110968

[68] Lan M, Zheng N, Dong X, Hua C, Ma H and Zhang X 2020 Bismuth-rich bismuth oxyiodide microspheres with abundant oxygen vacancies as an efficient photocatalyst for nitrogen fixation *Dalton Trans.* **49** 9123–9

[69] Wang Y, Lin L, Li F, Chen L, Chen D, Yang C and Huang M 2016 Enhanced photocatalytic bacteriostatic activity towards: *Escherichia coli* using 3D hierarchical microsphere BiOI/BiOBr under visible light irradiation *Photochem. Photobiol. Sci.* **15** 666–72

[70] Wei Z, Dong X, Zheng N, Wang Y, Zhang X and Ma H 2020 Novel visible-light irradiation niobium-doped BiOBr microspheres with enhanced photocatalytic performance *J. Mater. Sci.* **55** 16522–32

[71] Shan L W, He L Q, Suriyaprakash J and Yang L X 2016 Photoelectrochemical (PEC) water splitting of BiOI{001} nanosheets synthesized by a simple chemical transformation *J. Alloys Compd.* **665** 158–64

[72] Zeng L, Zhe F, Wang Y, Zhang Q, Zhao X, Hu X, Wu Y and He Y 2019 Preparation of interstitial carbon doped BiOI for enhanced performance in photocatalytic nitrogen fixation and methyl orange degradation *J. Colloid Interface Sci.* **539** 563–74

[73] Li H, Yang Z, Zhang J, Huang Y, Ji H and Tong Y 2017 Indium doped BiOI nanosheets: preparation, characterization and photocatalytic degradation activity *Appl. Surf. Sci.* **423** 1188–97

[74] Wang S-l, Wang L-l, Ma W-h, Johnson D M, Fang Y-f, Jia M-k and Huang Y-p 2015 Moderate valence band of bismuth oxyhalides (BiOXs, X = Cl, Br, I) for the best photocatalytic degradation efficiency of MC-LR *Chem. Eng. J.* **259** 410–6

[75] Li B, Shao L, Zhang B, Wang R, Zhu M and Gu X 2017 Understanding size-dependent properties of BiOCl nanosheets and exploring more catalysis *J. Colloid Interface Sci.* **505** 653–63

[76] Song J M, Mao C J, Niu H L, Shen Y H and Zhang S Y 2010 Hierarchical structured bismuth oxychlorides: self-assembly from nanoplates to nanoflowers via a solvothermal route and their photocatalytic properties *CrystEngComm.* **12** 3875–81

[77] Xiong J, Cheng G, Li G, Qin F and Chen R 2011 Well-crystallized square-like 2D BiOCl nanoplates: mannitol-assisted hydrothermal synthesis and improved visible-light-driven photocatalytic performance *RSC Adv.* **1** 1542–53

[78] Xu J, Meng W, Zhang Y, Li L and Guo C 2011 Photocatalytic degradation of tetrabromobisphenol A by mesoporous BiOBr: efficacy, products and pathway *Appl. Catal.* B **107** 355–62

[79] Feng Y, Li L, Li J, Wang J and Liu L 2011 Synthesis of mesoporous BiOBr 3D microspheres and their photodecomposition for toluene *J. Hazard. Mater.* **192** 538–44

[80] An H, Du Y, Wang T, Wang C, Hao W and Zhang J 2008 Photocatalytic properties of BiOX (X = Cl, Br, and I) *Rare Met.* **27** 243–50

[81] Hao H Y, Xu Y Y, Liu P and Zhang G Y 2015 BiOCl nanostructures with different morphologies: tunable synthesis and visible-light-driven photocatalytic properties *Chin. Chem. Lett.* **26** 133–6

[82] Gao X, Peng W, Tang G, Guo Q and Luo Y 2018 Highly efficient and visible-light-driven BiOCl for photocatalytic degradation of carbamazepine *J. Alloys Compd.* **757** 455–65

[83] Chang X, Huang J, Cheng C, Sui Q, Sha W, Ji G, Deng S and Yu G 2010 BiOX (X = Cl, Br, I) photocatalysts prepared using $NaBiO_3$ as the Bi source: characterization and catalytic performance *Catal. Commun.* **11** 460–4

[84] Henle J, Simon P, Frenzel A, Scholz S and Kaskel S 2007 Nanosized BiOX (X = Cl, Br, I) particles synthesized in reverse microemulsions *Chem. Mater.* **19** 366–73

[85] Ye L, Zan L, Tian L, Peng T and Zhang J 2011 The {001} facets-dependent high photoactivity of BiOCl nanosheets *Chem. Commun.* **47** 6951–3

[86] Cheng H, Huang B, Wang Z, Qin X, Zhang X and Dai Y 2011 One-pot miniemulsion-mediated route to BiOBr hollow microspheres with highly efficient photocatalytic activity *Chemistry* **17** 8039–43

[87] Zhang L, Cao X F, Chen X T and Xue Z L 2011 BiOBr hierarchical microspheres: microwave-assisted solvothermal synthesis, strong adsorption and excellent photocatalytic properties *J. Colloid Interface Sci.* **354** 630–6

[88] Xia J, Yin S, Li H, Xu H, Yan Y and Zhang Q 2011 Self-assembly and enhanced photocatalytic properties of BiOI hollow microspheres via a reactable ionic liquid *Langmuir* **27** 1200–6

[89] Li P *et al* 2020 Visible-light-driven nitrogen fixation catalyzed by Bi_5O_7Br nanostructures: enhanced performance by oxygen vacancies *J. Am. Chem. Soc.* **142** 12430–9

[90] Ye L, Tian L, Peng T and Zan L 2011 Synthesis of highly symmetrical BiOI single-crystal nanosheets and their {001} facet-dependent photoactivity *J. Mater. Chem.* **21** 12479–84

[91] Wang C, Zhang X, Yuan B, Shao C and Liu Y 2012 Simple route to self-assembled BiOCl networks photocatalyst from nanosheet with exposed (001) facet *Micro Nano Lett.* **7** 152–4

[92] Ding C, Ye L, Zhao Q, Zhong Z, Liu K, Xie H, Bao K, Zhang X and Huang Z 2016 Synthesis of $Bi_xO_yI_z$ from molecular precursor and selective photoreduction of CO_2 into CO *J. CO_2 Util.* **14** 135–42

[93] Bai Y, Ye L, Chen T, Wang L, Shi X, Zhang X and Chen D 2016 Facet-dependent photocatalytic N_2 fixation of bismuth-rich Bi_5O_7I nanosheets *ACS Appl. Mater. Interfaces* **8** 27661–8

Chapter 3

Efforts to improve the efficiency of BiOX photocatalysts

The utilization of photocatalysts is a highly efficient approach which may be employed for a multitude of purposes, including the destruction of diverse organic contaminants present in polluted water, the generation of hydrogen, the cleaning of air, and the exertion of antibacterial properties. In recent years there has been a significant surge in interest surrounding BiOX-based photocatalysts owing to their inexpensive nature as well as high yield in the fields of environmental management and energy conversion. The discernible physical and chemical characteristics of BiOX nanomaterials, specifically their energy band levels, and structures, along with their relaxed layered nanostructures, likely account for the enhancement of photocatalytic effectiveness in the presence of visible light. However, the actual usability of photoinduced electron and hole pairs is constrained by their high recombination rate. To address these limitations, researchers have devised several strategies, such as heterojunction formation, elemental doping, defect engineering, thickness adjustment, strain modulation, facet manipulation, solid solution formation, microstructure modification, bismuth-rich approaches, and compounding with carbonaceous materials. Taking into consideration the aforementioned statement, we have presented a comprehensive analysis of the current advancements in modified bismuth oxyhalide materials. Furthermore, this chapter also addresses the difficulties encountered in the development of efficient photocatalytic systems based on BiOX with enhanced performance. This chapter offers a distinct and comprehensive viewpoint on the development of BiOX-based photocatalyst materials, which have witnessed exponential expansion in research studies. The focus is on increasing the photocatalytic performance of these nanomaterials under visible light.

3.1 Heterojunction

The heterojunction has become a commonly used approach for enhancing a single semiconductor's photocatalytic effectiveness. This method can potentially broaden the wavelength of photoresponse for the system and accelerate the process of separating photoexcited charge carriers [1]. The band locations of the semiconductors allow for the classification of heterojunctions into three distinct categories, including Z-scheme heterojunctions, type III heterojunctions, type II heterojunctions, and type I heterojunctions (figures 3.1(a)–(d)). The type II heterojunctions, which are prevalent, exhibit the most favorable band positions for effective separation of electron and hole pairs, hence resulting in enhanced photocatalytic activity (PCA). Due to the advantageous energy configurations resulting from the relative positions of the conduction bands, the migration of photoelectrons from the conduction band of semiconductor A towards the conduction band of semiconductor B will occur. The valance band of semiconductor A receives holes at the same time that holes are transmitted from the valance band of semiconductor B. Because charge carriers are spatially separated from one another under the influence of light irradiation, the likelihood of recombination is reduced [2]. Peng *et al* [3] synthesized type II heterojunctions of $Bi_{12}O_{17}Br_2/Bi_{24}O_{31}Br_{10}$ using the calcination process of $BiOBr/Bi(OHC_2O_4).2H_2O$. The remarkable PCA that $Bi_{12}O_{17}Br_2/Bi_{24}O_{31}Br_{10}$ displayed for the breakdown of rhodamine B (RhB) and phenol has been assigned to the effective separation of charge carriers within the type II heterojunction. In brief, BiOX nanoparticles (NPs) have the potential to perform three crucial functions in semiconductor heterojunction systems. Beginning with a wide range of band redox potentials, BiOX can easily match the energy level of a variety of semiconductors. This matching ability provides a driving force for the directional photoinduced electron and hole charge separation. The next step is to exploit small band gap BiOX

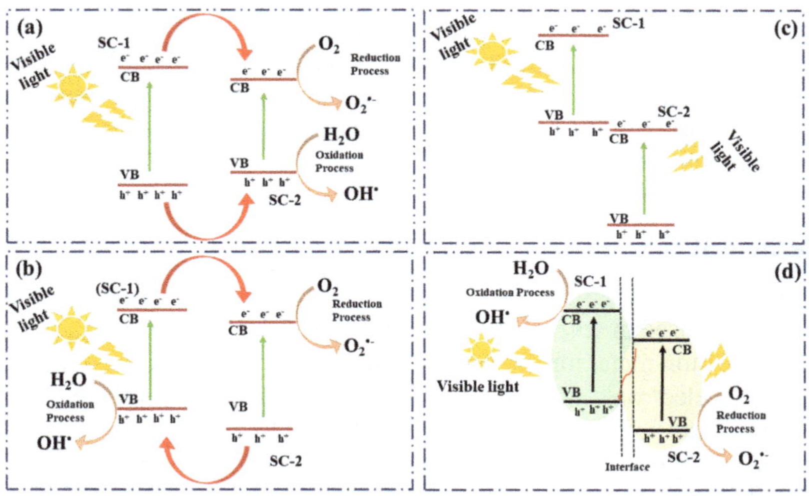

Figure 3.1. Schematic analysis of the (a) type I, (b) type II, (c) type III, and (d) Z-scheme mechanism. (Reproduced with permission from [9]. Copyright 2023 IOP Publishing.)

semiconductors (BiOBr and BiOI) to photosensitize other semiconductors to more efficiently consume solar energy. Finally, as a result of the structural variety of BiOX, carbon-based nanomaterials (NMs) may be readily connected with BiOX, which can somewhat boost the transfer and separation of photoexcited charge carriers. The PCA of BiOX has been improved in recent years using a variety of carbon compounds that have a wide surface area, good conductivity, and adsorption. In general, activated carbon and biochar often emphasized the growth of BiOX's adsorption ability. The surface of the BiOX contains reactive oxygen species (ROSs), which can more readily react with the macromolecular pollutants. When subjected to visible light (VL) irradiation, these pollutants usually disintegrate after becoming adsorbed in the dark. Other carbon materials, including carbon nanotubes (CNTs), carbon quantum dots (CQDs), and graphene (GO/RGO) possess superior conductivity and up-conversion. These properties facilitate the conversion of photoelectrons and expand the range of VL absorption [4]. Materials made of carbon often function as an 'electron trap' to capture photoelectrons. Moreover, CQDs have the capability to significantly contribute to the phenomenon of transforming near-infrared light into ultraviolet light. For instance, Xia *et al* [5] conducted an in-depth investigation of the impact that CQDs have on improving the PCA of BiOBr. The remarkable enhancement in the stability and photodegradation efficiency of ciprofloxacin (CIP) may be attributed to the high separation efficiency of photoexcited electron/hole charge pairs as well as the broad light adsorption properties exhibited by CQDs. In particular, BiOBr/graphene, BiOI/graphene, and BiOCl/graphene have all been employed as a result of the strengthened link that exists between BiOX and graphene [6, 7]. In addition, graphene may be utilized as a co-catalyst to capture electrons created by light. Due to this, the electrons that transfer to graphene from BiOX can effectively block the recombination of photoexcited charge carriers, prolong the carrier lifespan, and increase the PCA. Additionally, the PCA was significantly enhanced after g-C_3N_4 was coupled with BiOI [8]. Both BiOI and g-C_3N_4 were excited when they were subjected to VL irradiation. The photogenerated electron on the g-C_3N_4 conduction band goes to the BiOI conduction band, although the photogenerated hole on the BiOI valance band moves to the g-C_3N_4 valance band. Effective charge separation takes place, which increases the PCA.

3.2 Elemental doping

In structural engineering, elemental doping is recognized as an essential method. Doping has been used in several investigations to regulate the photocatalytic activity of BiOX. Transition metal ion inclusion into lattices of BiOX for the goal of creating or changing the electronic states has received greater attention in the past few years. Furthermore, it can result in imperfections in the BiOX lattice or have an impact on the substance crystallinity, both of which serve to prevent the pairing of holes and electrons, which results in the production of carriers that are stable over time. The performance of these materials' photoelectrochemical (PEC) capabilities can be significantly impacted by doping with specific metal ions, which may also improve the semiconductors' light absorption area. The introduction of transition metal ions through doping can give rise to the creation of impurity energy levels, potentially causing a shifting of the energy band towards a longer wavelength. The observed shift could be assigned to the electron passage from impurity bands to either the conduction band or valence band. The following section provides an overview of the outcomes that occurred as a consequence of doping BiOX-based materials with rare earth metals, transition metals, non-metals, and noble metals.

3.2.1 Noble metal modification

One effective method for enhancing the process of photocharge separation involves the doping of noble metals onto the BiOX surface, thereby inducing the formation of a space-charge dividing region, commonly referred to as the Schottky barrier. The interaction between the noble metal and BiOX leads to a modification in the Fermi energy levels. This occurs as an electron transfers from a BiOX atom possessing an elevated Fermi level to a metal atom with a smaller Fermi level. The metal and semiconductor are each overcharged with negative and positive charges due to the Schottky barrier's presence. A Schottky barrier in photocatalysis also functions as an effective electron trap that hinders the reconnection of the hole and electron, which frequently leads to an improved capacity for photocatalysis. Li *et al* [10] manufactured Pd/BiOBr using the methodology of photo deposition [10]. The results of their study demonstrated that the introduction of palladium (Pd) triggered the formation of additional oxygen vacancies (OVs) over the BiOBr surface. This occurrence can be attributed to the interaction of electrons transpiring on the interface between BiOBr and Pd. The facilitation of oxygen, as well as toluene molecule adsorption, is enhanced by the coexistence of Pd and OVs on BiOBr. Because of this, the toluene and oxygen are activated at the Pd/BiOBr interface, which also controls charge division. The PCA of BiOBr is enhanced by a factor of 1.5, demonstrating a notable efficiency in the selectivity of toluene oxidation to benzaldehyde, achieving a 99% yield. The study done by Yu *et al* [11] aimed to investigate the influence of noble metal deposition, specifically palladium (Pd), platinum (Pt), and rhodium (Rh), on the PCA of BiOX. Under ultraviolet light, the effects of the noble metals are ranked as follows: Pt > Pd > Rh, whereas under VL, the order is Rh > Pt > Pd. The noble metal NPs also serve as electron traps, which accelerates the electron and hole distribution and, as a result, improves the VL activity. Arumugam *et al* [12] recently produced noble metals (Au, Pt, Pd, and Ag) doped BiOBr to increase the photodegradation efficacy of phenol (figure 3.2). The experimental findings demonstrated that the efficacy of phenol breakdown is significantly enhanced using Pd-doped BiOBr compared to pure BiOBr. This improvement in activity was due to the surface plasmon resonance's facilitation of increased surface area, excellent photon absorption, and a decreased recombination rate. In addition, the BiOBr sample doped with Pd exhibited excellent stability and reusability after being put through three cycles.

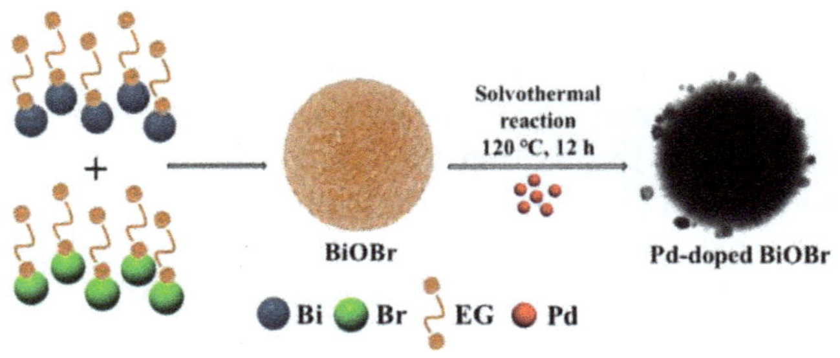

Figure 3.2. Illustration of the preparation of Pd-doped BiOBr material. (Reproduced with permission from [12]. Copyright 2023 Elsevier.)

3.2.2 Transition metal/metal doping

The introduction of dopants containing transition metals may be employed to induce lattice defects or modify the crystallinity of a semiconductor. This approach effectively stops the photoexcited charges from recombining, consequently accelerating the lifetimes of these charges. The introduction of certain metals through doping may potentially enhance the light absorption range in semiconductors. This effect is in addition to the impact of transition metals' diverse valence states and 3d orbitals, which exert significant effects on the semiconductor's photoelectrochemical qualities. An energy band shift towards a higher wavelength might be caused by doping with specific transition metals, such as Mn^{3+} and Fe^{3+}. The addition of metals over the crystal lattice results in impurity levels falling inside a band that a semiconductor is not allowed to have. The change from CB or VB impurity levels to lower impurity levels is thought to be the cause of the redshift. In general, highly effective doped photocatalysts depend on tuning by doping with transition metals for a significant portion of their efficiency. This doping should meet two conditions, which are as follows: (i) The dopant is capable of capturing both the electron and the hole, which enables efficient local separation, and (ii) the electron and hole that were previously imprisoned can be freed and will travel to the interface. A wide variety of transition metal ions, including copper (Cu), zinc (Zn), titanium (Ti), iron (Fe), aluminum (Al), tin (Sn), and manganese (Mn), have been utilized as dopants in BiOX materials. Doping BiOCl with a trace amount of manganese, in conjunction with OVs, has the potential to reduce the band gap of the material as well as increase its optical absorption in the visible and infrared spectra [13]. After doping with tungsten, the band gap was reduced, and experimental measurements confirmed the creation of an impurity band in the forbidden area of the BiOCl crystal structure. The reduction in band gap was shown by theoretical calculations based on first-principles theory [14]. For instance, iron and copper might be introduced into the BiOCl system employing a process involving the self-doped reactive ionic liquids 1-octyl-3-methylimidazolium tetrachloroferrate ([Omim]$FeCl_4$) as well as [Omim] $CuCl_3$ [15]. In this methodology, the second reactant, known as [Omim]$CuCl_3$, served as a model for Cu/BiOCl microspheres (MSs) and a source of Cu and Cl. Most crucially, Jiang *et al* produced an Ag and Ti-doped BiOBr photocatalyst, which resulted in the effective augmentation for dye degradation [16]. In a separate investigation, Chen *et al* synthesized a nanosheet composed of BiOBr doped with Ni, Mo, and Fe [17]. This particular work aimed to explore the potential of the synthesized material for nitrogen fixation under VL conditions. The utilization of transition metal doping in conjunction with the presence of OVs leads to the generation of modulated band structures, enhanced separation of charge carrier separation, and facilitated electron transfer to N_2. Consequently, these factors collectively contribute to enhancing N_2 activation capacity and adsorption in modified BiOBr NSs. The Fe-doped BiOBr containing OVs demonstrates superior performance in the synthesis of ammonia (NH_3), with a PCA of 46.1 μmol g^{-1} h^{-1}, surpassing that of pristine BiOBr by a factor of six. Notably, this enhanced catalytic activity is achieved without using any sacrificial agent. The findings will throw new insight into charge carrier separation, activation sites, and rational engineering of adsorption to build effective N_2 fixation photocatalysts.

3.2.3 Non-metal doping

The incorporation of non-metal dopants results in the creation of supplementary extrinsic electronic states inside the band gap, leading to notable effects on the material's charge carrier confinement and light absorption characteristics. The prevailing hypothesis posits that the upward shift of the valence band to its highest level occurs as a result of the hybridization between local states and valence orbitals [18]. Iodine (I), sulfur (S), nitrogen (N), carbon (C), and boron (B) are all examples of dopants that have been described as being suitable for use in the production of non-metal-doped BiOI, BiOCl, and BiOBr. Boron has a relatively tiny ion radius, which is around 0.023 nm. In the crystal structure of BiOX, the dispersion of B is made easier by the material's low weight and semi-conductivity. Using borax as the precursor of B, Yu *et al* [19] utilized a solvothermal technique for the preparation of B-doped BiOCl photocatalysts. The introduction of B-doping leads to a notable augmentation in the surface-area-to-volume ratio.

In addition to this, the (001) exposed facets in the B-doped BiOCl nanosheets (NSs) increased the efficiency with which charges could be separated. Because of this, the optimized B1.0-BiOCl demonstrated a strong PCA for the destruction of bisphenol A, phenol, and RhB. Boron-doped BiOBr NSs were created by Wang and colleagues [20], and in these NSs, the B dopant generates an impurity level that speeds up charge separation. B-doped BiOBr photocatalysts with hierarchical MSs were produced by Liu and colleagues [21] for degrading phenol and RhB. The addition of B dopant in the lattice of BiOBr occurred by a substitutional mechanism. Consequently, the resulting BiOBr material that was doped with B exhibited surfaces that were enriched with hydroxyl groups. The process enhanced the adsorption of dye molecules, hence boosting the effectiveness of separating photoexcited charge carriers. P- and B-doping in BiOBr single-layer has been observed by first-principles computation [22]. The addition of P-doping effectively resulted in the reduction of the band gap of BiOBr. Conversely, the incorporation of B-doping led to an expansion of the absorption range for VL and a decrease in the recombination rate of photoinduced charge carriers. Based on the results

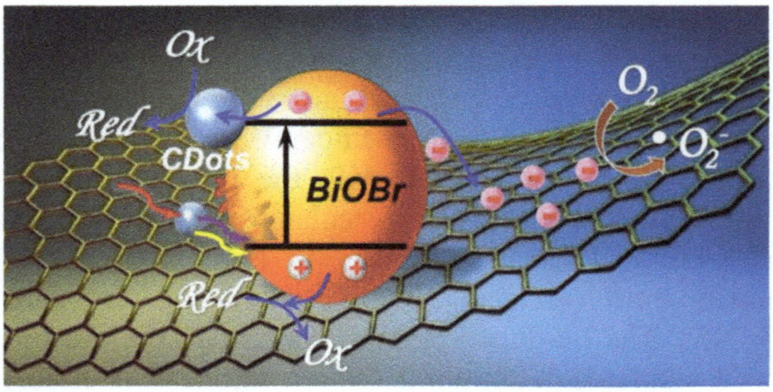

Figure 3.3. The mechanism of the transfer and separation of photoinduced charges in the GO/CDots/BiOBr system under VL irradiation. (Reproduced with permission from [24]. Copyright 2018 Elsevier.)

obtained from the calculations, it can be inferred that the incorporation of B-doping led to a notable increment in the quantity of catalytic active sites. Hence, the efficacy of B-doped BiOBr has been demonstrated as a feasible approach for enhancing photodegradation efficiency. Carbon doping has been observed to enhance the efficiency of charge transmission as well as separation in photoinduced systems. In this regard, Zeng *et al* [23] used glucose as a source of carbon to produce C-doped BiOI (C–BiOI) photocatalysts. Introducing doped carbon clusters into the BiOI interlayer structure resulted in several notable effects. These include modifications to the lattice periodicity, the creation of vacancies, and an augmentation in the particular area of the surface. These changes collectively led to a reduction in the band gap as well as an enhancement in the separation of charge efficiency. In the context of the decolorization of methyl orange (MO), the improved C–BiOI exhibited a degradation rate that was 4.44 times greater compared to pristine BiOI. Composites consisting of carbon nanodots (CDots) and graphene oxide (GO) that have been co-doped with BiOBr have been produced by Qu *et al* [24]. As the electron reservoir as well as the electron transport center, GO reduced the recombining rate of photoexcited charges in this ternary composite. The CDots significantly improved the capacity of BiOBr to absorb and make use of VL. Under the influence of VL, the GO/CDots/BiOBr combination exhibited the best PCA against the breakdown of 4-chlorophenol (4-CP) (figure 3.3). Furthermore, a solvothermal method has been used to decorate N-doped BiOBr NSs, which were subsequently put onto the carbon fiber (CFs) surface [25]. The remarkable PCA that 3D N-doped BiOBr/CFs displayed for the RhB and methanol degradation was mostly caused by the significant absorption of VL from all directions.

In their study, Jiang *et al* [26] used a solvothermal preparation method to manufacture photocatalysts consisting of BiOCl–S. The separation of photogenerated pairs of electrons and holes was made quicker by a uniform distribution of S elements on BiOCl, which broadened the light absorption range. In addition, the BiOBr band structure can be manipulated by S doping to obtain the desired results. Wang *et al* [27] produced 10 nm thick S-doped BiOBr using the process of hydrothermal synthesis. The BiOBr band gap was decreased thanks to doping with S, which also made it possible for BiOBr to have a strong reaction to VL. In addition, the efficiency for the separation of charge was increased by using S doping. As a consequence of this, optimized S0.2-BiOBr NSs demonstrated the maximum activity for the breakdown of bisphenol A when subjected to the irradiation of VL. I-doped BiOCl MSs were made by Zhang *et al* [28] using a procedure that involved the assistance of a microwave. In addition to decreasing the band gap and promoting the photogenerated carrier's separation, I doping enhanced a particular area of the surface of BiOCl. As a result, BiOCl doped with I produced a great clearance ratio of 91.2% in 60 min when used for 2,4-dihydroxybenzoate degradation.

3.2.4 Rare earth metal doping

Doping of rare earth metals improves electron–hole pair separation and photocatalysis. Oxides of rare earth metals have a variety of crystal forms, which enables them to have excellent conductivity, excellent thermal stability, and extremely selective adsorption. Rare earth metals have many electronic energy levels because they have a 4f electron shell. These levels act as places for photocharge carriers to trap electrons. A solvothermal method produced Y-doped MSs of BiOBr [29]. When the Y was present, PCA for the breakdown of CIP, as well as RhB under VL irradiation, was improved. The creation of a sub-band underneath the conduction band, in which electrons can become trapped and so reduce the recombination of photochagre (figure 3.4), is responsible for the improvement in the PCA of the material. Similarly, the breakdown of tetracycline was enhanced by the visible light PCA of Y-doped BiOCl [30]. Dopants for the BiOX system have also recently been derived from other rare earth elements, including erbium and europium [31]. Dash *et al* designed and manufactured a photocatalyst made up of poly(vinyl alcohol)-functionalized Eu-doped BiOX nanoflakes. This photocatalyst displays dramatically

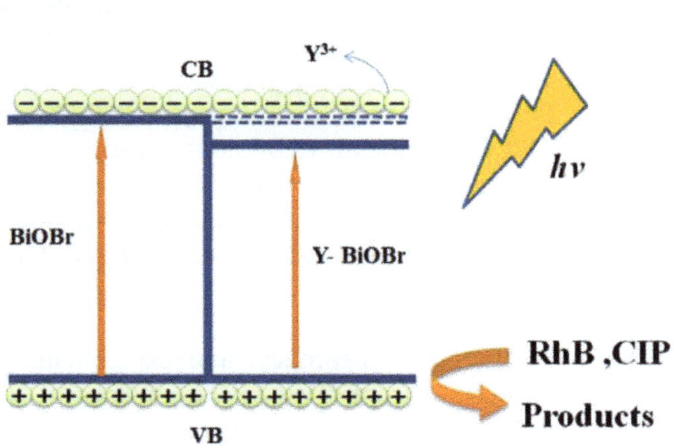

Figure 3.4. Schematic mechanism of the photocatalytic activity by yttrium-doped BiOBr. (Reproduced with permission from [29]. Copyright 2015 Elsevier.)

improved RhB degrading activity [32]. In the La^{3+}-doped BiOBr MSs that were synthesized by utilizing an ionic liquid ([C16mim] Br)-assisted solvothermal technique [33] the replacement of Bi^{3+} by La^{3+} was examined by XRD peak shift. However, neither La_2O_3 nor another phase of La was identified. The enhanced photodegradation of CIP was seen in La-doped BiOBr due to the efficient trapping of photoelectrons by La, which greatly inhibits photocharges from recombining.

3.3 Defect engineering

Vacancies and defects can modify the charge distribution and electrical energy levels of semiconductors, influencing their photocatalytic efficacy. Surface imperfections function as sites for the adsorption of reactants and for the trapping of photogenerated charge carriers. Introducing defects in BiOX enables the material to have its micro- and electron structure, atomic coordination number, and charge transfer modified. The prevalence of oxygen atoms within the structure of numerous Bi-based materials makes OVs the most commonly utilized sort of defect. Methods such as solvothermal [34–37], illumination [38–40], surfactant-assisted synthesis [41–43], and electrochemical reduction [44] may all be utilized to induce oxygen defects in BiOX.

Among these approaches, the solvothermal method is the one that is utilized the most frequently to induce an oxygen deficiency in BiOX by utilizing ethylene glycol (EG) as the solvent [36, 37]. EG has the ability to undergo reduction under conditions of high pressure and temperature, and it can help create OVs by partially reducing the Bi^{3+} component of BiOX. Li *et al* used an EG-assisted solvothermal approach to synthesize the exposed {001} facet of BiOBr using OVs and then used it to convert atmospheric N_2 to NH_3 [34]. When the sample that did not contain OVs was subjected to further sintering in an O_2 environment, the color changed to white. NH_3 was shown to be present across BiOBr that had OVs, but not in BiOBr that did not contain OVs. This phenomenon arose owing to the oxygen vacancy in BiOBr, which acted as the catalytic activation site and exhibited the capability to chemisorb N_2 onto its surface (figures 3.5(a) and (b)). By using a one-step solvothermal process, Wang *et al* could generate OVs that were rich in S doping in BiOBr. The BB-5S catalyst showed a remarkable ability to decompose 98% of the 4-chlorophenol (4-CP) over 2 h under VL. This performance was much superior to that of the OV-poor sulfur-doped BiOBr catalyst, exhibiting an 18.0-fold increase, as well as the pristine ultrathin BiOBr catalyst, which showed a 4.9-fold increase in efficiency. DFT computations and experiments indicated that a sub-band was created under the synergy of S doping and OVs. This sub-band significantly improved the VL absorption capabilities of the material, hindered photocharge recombination, and contributed to an increase in PCA [45]. Cui *et al* [46] used a solvothermal method to manufacture OVs-rich BiOCl NSs. Theoretical and experimental evidence supported a new energy level derived from OVs. In the presence of VL, BiOCl NSs doped with OV were able to generate oxygen photocatalytically (figures 3.5(c) and (d)). Moreover, BiOCl with a high oxygen vacancy content exhibited enhanced VL photocurrent, as well as improved efficiency in photocharge migration and separation, when compared to BiOCl with a lower number of OVs.

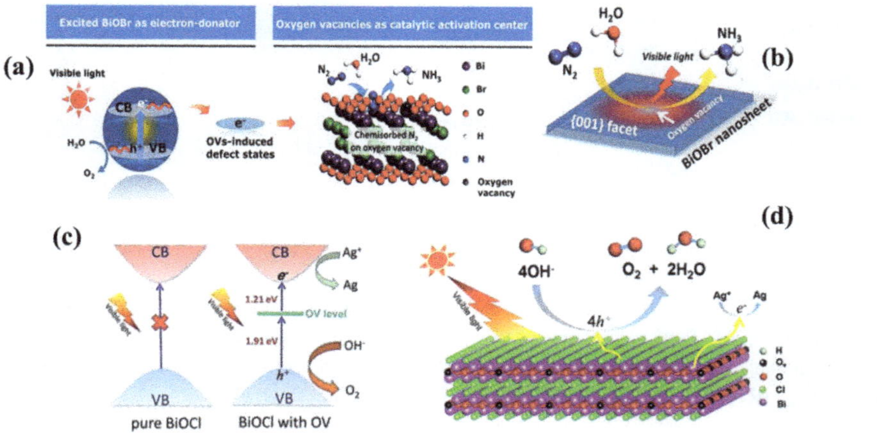

Figure 3.5. (a) Schematic illustration of N_2 fixation by BiOBr-OV. (b) N_2 fixation mechanism with water as both the proton and solvent source. (Reproduced with permission from [34]. Copyright 2015 American Chemical Society.) (c) Photooxidation mechanism and (d) schematic illustration of O_2 evolution over BiOCl-OVs in the presence of VL irradiation. (Adapted with permission from [46]. Copyright 2018 The Royal Society of Chemistry.)

The utilization of surfactant-assisted fabrication has emerged as a contemporary approach for the production of BiOX materials that exhibit a high abundance of OVs [42]. Polyvinylpyrrolidone (PVP) is the surfactant that is most frequently utilized. The positively charged Bi^{3+} in BiOX readily interacts with the negatively charged C = O in the framework of PVP to form BiOX. BiOX's surface energy is reduced as a result of PVP's adsorbed presence on its surface, which is effective for the production of OVs [42, 43]. For instance, Xue *et al* [43] utilized the PVP-assisted synthesis approach to produce BiOBr that was abundant in OVs (VO-BiOBr) and then utilized it in the photocatalytic synthesis of ammonia (NH_3). VO-BiOBr exhibited a much higher capacity for the adsorption of nitrogen dioxide than did BiOBr. The photocatalytic rate of NH_3 production by BiOBr and VO-BiOBr, on average, were 5.75 and 54.70 μmol g^{-1} h^{-1}, correspondingly. As a result, it was shown that the insertion of OVs promoted both the adsorption of nitrogen molecules as well as the photocatalytic generation of NH_3. The concentration of OVs produced by this approach may be modified to the desired level by altering the amount of PVP used.

A beneficial influence on the photocatalytic performances of BiOX can be attributed to the addition of an adequate quantity of OVs. Using a straightforward hydrothermal technique, Chen *et al* [47] were able to fabricate nanoplates of BiOCl with {001} exposed active crystal faces. They discovered that the synthesized sample exhibited a rate constant (*k*) for the breakdown of RhB in the presence of VL irradiation that was about three times higher (0.034 min^{-1}) compared to that observed in the presence of UV–visible irradiation (0.012 min^{-1}). This is because OVs may be easily produced during UV light irradiation. Under the influence of VL, the primary mechanism by which RhB is broken down by BiOCl NSs is the

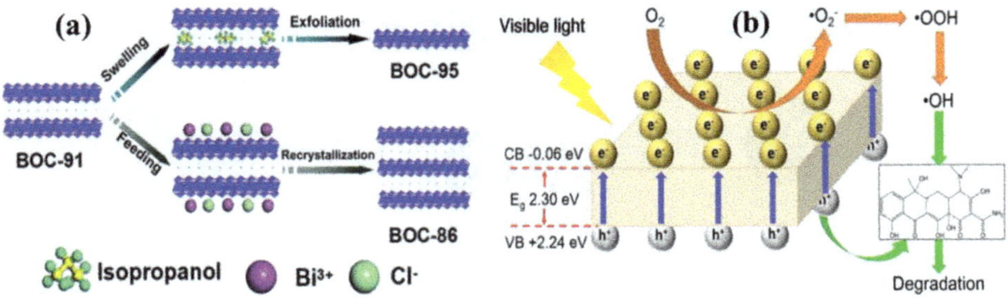

Figure 3.6. (a) Schematics of the hydrothermal-induced recrystallization and liquid exfoliation methodologies for tuning the {001} facet percentages of Bi_3O_4Cl NSs. (Adapted with permission from [50]. Copyright 2014 The Royal Society of Chemistry.) (b) Schematic of photocatalytic tetracycline hydrochloride degradation over $Bi_{24}O_{31}Br_{10}$. (Reproduced with permission from [51]. Copyright 2017 Elsevier.)

photosensitization process. In contrast, OVs are produced on {001} facets of BiOCl NSs during UV–vis light irradiation and trap photoexcited electrons on the conduction band, preventing the production of $^{\bullet}O_2^-$ radicals. As a result, the oxygen deficiency in the NSs of BiOCl present here lowers the effectiveness of photocatalytic RhB breakdown.

3.4 Thickness tuning

Thickness tuning is a further method that may be utilized to improve the PCA of BiOX. Whenever the thickness of the BiOX is decreased, it is possible to achieve quantum confinement to adjust the band edge structure, that is advantageous for meeting the potential requirements of many kinds of catalytic procedures. Additionally, the distance of charge migration from the interior to the surface may be shortened, increasing the effectiveness of charge separation. The active sites needed for catalytic processes can be provided by atoms with low coordination numbers and dangling bonds [48, 49]. Li *et al* demonstrated the thickness-dependent PCA toward the degradation of salicylic acid. They produced thicker and thinner Bi_3O_4Cl by liquid phase exfoliation under further recrystallization or isopropanol [50]. The Bi_3O_4Cl with the lowest thickness has the most photodegradation activity because it has the largest internal electric field (figure 3.6(a)). The preparation of $Bi_{24}O_{31}Br_{10}$ with a thinner thickness was achieved by employing a greater quantity of NH_4Br. Thinner $Bi_{24}O_{31}Br_{10}$ NSs can potentially achieve enhanced separation of charge efficiency, resulting in a higher concentration of reactive OVs, which promotes the degradation of tetracycline hydrochloride (figure 3.6(b)) [51].

3.5 Strain modulation

The 2D stacked structure of BiOX materials makes them very sensitive to changes in strain when they are synthesized [52]. In addition, the BiOX electronic structure is susceptible to significant modification even when there is only a little alteration in the inner strain. This was something that Du and colleagues looked into, and they discovered how the strain impacts the PCA over a BiOBr circle and BiOBr square [53].

They began by observing certain in-plane creases in the BiOBr square using HRTEM pictures. Subsequently, using GPA simulation, they derived an inhomogeneous strain distribution from the material [52, 54]. Their study shows unequivocally that the specific type of local lattice distortion is exclusively localized on the BiOBr square. In contrast, it has been observed that the strain distribution on the surface of the BiOBr circle is homogeneous, indicating a much lower degree of lattice distortion compared to that observed in the BiOBr square. In general, it was evident that the level of strain in the square-shaped BiOBr was significantly higher compared to the circular-shaped BiOBr. During this time, the DFT calculations were used to conduct an in-depth study of the strain influence on the electronic structure [55]. By lowering the size of the band gap in the photocatalysts, two different kinds of strains, known as tensile strain and compressive strain, might increase the photocatalysts' capacity to collect solar light over a broad range.

3.6 Facet control

The photocatalytic process on BiOX photocatalytic NMs mostly occurs on their surface, and the crystal facets that are visible on this surface have a substantial impact on the overall efficiency of the photocatalytic material. This is a fundamental property of crystalline materials. As a result of the distinct geometric and electrical structures that each facet possesses, the photocatalytic performance of the various facets is distinct [56, 57]. In general, the crystal facet that has a greater surface energy also has a higher level of photoactivity. However, it tends to vanish before the crystal formation process even begins [58]. Capping reagents are required to disclose the crystal facet that possesses a greater surface energy [59]. Surfactants and H^+ are two types of capping reagents that are frequently utilized in the process of crystal facet tailoring using BiOX [60, 61]. For example, Jiang *et al* [61] utilized a one-step hydrothermal methodology to manufacture BiOCl NSs with highly exposed {001} and {010} facets. The aqueous solution of $Bi(NO_3)_3{\cdot}5H_2O$ undergoes a reaction with water, forming $BiONO_3$ and H^+ ions. Following this, Cl^- formed Cl–Bi–O–Bi–Cl slices by reacting with $BiONO_3$. Because of the strong binding relationship between terminated oxygen and H^+, the H^+ was adsorbed on the {001} facet. Furthermore, this interaction hindered the progress of crystal facet growth, resulting in the development of BiOCl with a prominently exposed (001) facet, denoted as BOC-001. In addition, the BiOCl (BOC-010) crystal with a prominent {010} facet was synthesized under a pH of 6. The photocurrent density of BOC-001 was effectively greater when compared with BOC-010 due to the more favorable transfer and separation of charge along the [001] direction, as a result of the internal electric field, during UV light irradiation. When subjected to UV light irradiation, the apparent reaction rate constant normalized by the surface area (k') and the apparent reaction rate constant (k) displayed by BOC-001 were both found to be greater than those exhibited by BOC-010. In spite of this, the normalized k' values of compound BOC-001 as well as compound BOC-010 become quite similar when exposed to VL. In addition, BiOBr that had high exposed crystal facets (102) and (001) was also produced and utilized for the catalytic disintegration of RhB [62]. Because it has a shorter band gap and a wider light absorption range, the (102) facet dominant form of BiOBr exhibited a greater degree of PCA when subjected to illumination by VL. Solvothermal synthesis was another way that was used to produce BiOI with highly exposed crystal facets (110) and (001) [63]. The photocatalytic breakdown rate of bisphenol A (BPA) was higher when BiOI with a prominently exposed (110) crystal facet was utilized and subjected to VL irradiation. The BiOI-110 demonstrated a higher degree of PCA during the breakdown of BPA due to the lower adsorption energy of oxygen on its (110) facets (0.209 eV) compared to its (001) facets (0.656 eV). The findings demonstrated that BiOI-110 could produce a greater quantity of $^{\bullet}O_2^-$, which resulted in an increase in the PCA under VL. As a consequence of this, it can trap a greater number of electrons, which increases the efficiency with which photoinduced electrons and holes may be separated, leading to the production of a greater number of ROSs that are associated with electrons and holes [63]. The BiOI-110 sample is capable of generating $^{\bullet}OH$ due to its comparatively larger oxidation potential concerning BiOI-001. This is due to the fact that the oxidation potential of BiOI-110 is higher than that of BiOI-001. In conclusion, if the reaction solution's pH value is tuned appropriately and reasonably, it is possible to synthesize a BiOX-exposed particular crystal facet. Numerous prior studies have examined the impact of crystal facets on the PCA of BiOX NMs. Nonetheless, the task of generating crystal facets with a high level of reactivity and comprehensively elucidating the fundamental reaction mechanisms occurring on these specific facets remains a formidable challenge [64]. In the synthesis of the material, Wu *et al* [65] utilized a straightforward chemical precipitation technique to generate tetragonal BiOBr, wherein a significant presence of either {010} or {001} crystal faces was achieved. Based on the findings derived from scanning electron microscopy, it was observed that BiOBr particles exhibiting a prevalence of the {010} facet revealed a larger plate-like morphology. In comparison, those dominated by the {001} facet displayed small in size, more irregular particles (figure 3.7). Mott–Schottky and ultraviolet visible diffuse reflectance spectra analysis showed that BiOBr-010 had a band gap and valence band potential of 2.95 and 2.71 eV versus a normal hydrogen electrode (NHE), respectively, while BOBr-001 showed 3.15 and 2.63 eV. The superiority of BiOBr-010 over BOBr-001 in terms of photo-oxidative capability was proven in the aqueous phase for the breakdown of formic acid and water oxidation.

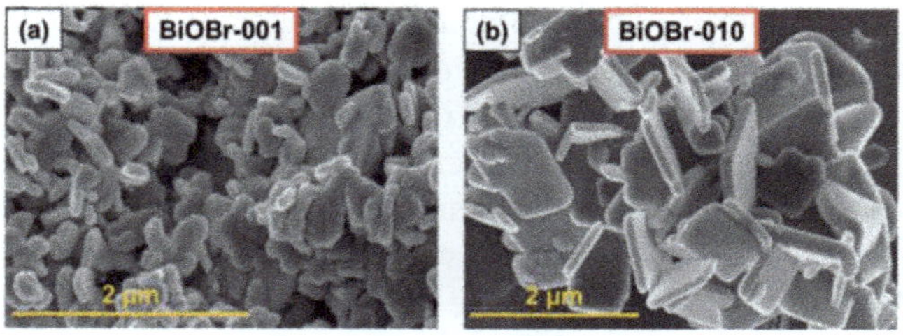

Figure 3.7. SEM micrographs (a) BiOBr-001 and (b) BiOBr-010. (Adapted with permission from [65]. Copyright 2017 The Royal Society of Chemistry.)

The prevention of photogenerated electron and hole recombination is what is responsible for the increased photo-oxidative capacity of BiOBr-010. Furthermore, it was observed that the increased efficiency of charge transfer and the decreased resistance of charge transfer present in BiOBr-010 were advantageous for improving photoelectrochemical (PEC) performance. These findings explain why BiOBr-010 has a higher current density and stronger photo-oxidative activity while having a lower specific surface area. They also demonstrate how crystal facet engineering may be used to improve the photocatalytic efficacy of a material.

3.7 Solid solution

The construction of solid solutions allows for the band gap structure, electrical structure, as well as crystal structure of photocatalysts to be tuned. Because solid solutions can contain a wide variety of constituents, it is essential to use a solid photocatalyst with enhanced activity. Solid solutions of BiOX have been shown to exhibit potent PCA, according to reports [66].

The most frequent source of Bi is $Bi(NO_3)_3{\cdot}5H_2O$, while sources of halogen include halogen-containing surfactants, KX, and NH_4X. A large number of solid solutions based on BiOX have been produced, including $BiOCl_xBr_{1-x}$ [66], $BiOCl_xI_{1-x}$ [67], and $BiOBr_xI_{1-x}$ [68]. Methods of synthesis include the electrodeposition technique [69], hydrothermal method [66, 70], solvothermal method [71, 72], and room temperature precipitation method [67, 73]. The utilization of precipitation at room temperature is greatly employed in synthesizing BiOX solid solutions due to its advantageous features, including convenient operation, simple batch preparation, and mild reaction conditions [67]. The synthesis of $BiOCl_xI_{1-x}$ solid solution was carried out at room temperature by Zhang *et al* [67], who employed EG and water as their solvents. Based on the results of a photocurrent test, the photocurrent intensity of $BiOCl_{0.9}I_{0.1}$ was significantly higher when compared to BiOI and BiOCl. This indicates that the development of a solid solution facilitated the carrier's separation. The utilization of XPS and XRD techniques provided evidence that introducing the doped iodide ion effectively substituted the chloride ion within the crystalline structure of BiOCl. This substitution led to the creation of a homogeneous solid solution denoted as $BiOCl_xI_{1-x}$, rather than a mere mixture of BiOI and BiOCl. Band gap values of 1.622, 2.261, and 2.528 eV, respectively, were determined to be associated with the indirect band gap semiconductor properties of BiOI, BiOCl, and $BiOCl_{0.9}I_{0.1}$ based on their respective electronic band structures. To examine the photodegradation potential of the samples, the decomposition of RhB was utilized. According to the findings of the experiment, $BiOCl_{0.9}I_{0.1}$ exhibited the highest PCA since the rate of degradation it catalyzed was the quickest. In addition, Sasson and colleagues [74] discovered that when the surfactant cetyltrimethylammonium halide was used as a source of halogen and a structure-directing agent, the solid solution of $BiOCl_xBr_{1-x}$ revealed a three-dimensional microsphere shape. The solid solution that was formed had a two-dimensional sheet structure when KBr and NaCl were utilized as the source of the halogen. As a result, it can be concluded that sources of halogen have an impact on the microstructure as well as the physical and chemical characteristics of BiOX solid solutions.

For the manufacture of solid solutions of BiOX, two methods that are often utilized are the precipitation approach at room temperature and the hydrothermal method [68, 75, 76]. For example, Pan *et al* [70] used KI and NaCl as their source of halogen to generate a sheet-shaped $BiOCl_{1-x}I_x$ solid solution using the hydrothermal process. The RhB degradation experiment revealed that the PCA of the solid solution was noticeably heightened in comparison to that of the pure BiOX. The solid solution of $BiOCl_xBr_{1-x}$ was synthesized by Yu *et al* [66] using the hydrothermal technique, with NH_4Cl and NH_4Br serving as the halogen sources. They explored how the molar ratio of halogen ions affected the band gap value and the shape of the material. Due to the synergistic impact of an exposed particular facet as well as an optimized band structure, the $BiOCl_{0.5}Br_{0.5}$ solid solution was shown to have the maximum PCA in the experiment that measured degradation. In addition, the solvothermal approach is the method that is utilized the majority of the time for the synthesis of three-dimensional MS solid solutions containing BiOX [71]. The formation and nucleation of crystals in solvothermal and hydrothermal processes are distinct. This is because of the disparities in viscosity and saturation vapor pressure between water and EG. As a consequence, the samples' physical and chemical characteristics are distinct from one another. In addition to that, electrodeposition is also utilized during the fabrication of solid solutions of BiOX. For example, Yu *et al* [69] used p-benzoquinone and nitric acid to facilitate the electrodeposition of a solid solution of $BiOI_{0.5}Cl_{0.5}$ onto the surface of indium tin oxide (ITO). A PEC biofuel cell was developed using the modified electrode as the photocathode. This construction widened the use of solid solutions of BiOX greatly.

In a nutshell, solid solutions of BiOX may be made by employing $Bi(NO_3)_3.5H_2O$ as a source of Bi and NH_4X, and KX as sources of halogen. These solid solutions can be made using the precipitation, electrodeposition, solvothermal, and hydrothermal methods. The band gap value in solid solutions containing BiOX can be manipulated by adjusting the halogen ion ratio, while the enhancement of the PCA in these solutions can be achieved by adjusting the electronic structure. Solid solutions have the potential to enhance their photocatalytic properties by modifying their energy band gap which can be achieved by elevating the valence band position and lowering the conduction band position. The band construction has an effect on the bandwidth. In the presence of sufficient excitation energy, photogenerated electrons can penetrate the conduction band at any location. Another benefit of solid solution photocatalysts is their easy tunability of band gaps, which allows for optimization of the PCA of the materials. Charge carriers recombination is reduced owing to the alloying effects caused by BiOX in solid solutions.

3.8 Microstructure modulation

Microstructure, such as dimensionality, specific surface area, shape, and size, strongly affects materials' physicochemical and chemical characteristics [77, 78].

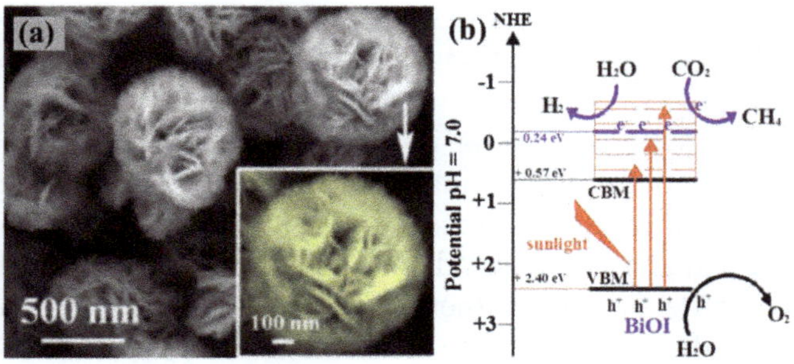

Figure 3.8. (a) SEM micrographs of BiOI flower-like hierarchical structure. (b) Proposed mechanism for photocatalytic reduction of CO_2 in the presence of sunlight irradiation. (Reproduced with permission from [79]. Copyright 2014 Elsevier.)

BiOX NMs include 1D nanofibers, 2D NS/nanoplates, and 3D MS/hierarchical nanostructures [79–82]. More attention has been generated by the 3D assemblies due to the unique architecture and outstanding PCA of these structures. Zhang *et al* [79] produced a BiOI photocatalyst with a hierarchical structure resembling a flower at room temperature using the direct hydrolysis approach. (figure 3.8(a)) They then used it for the first time for the photoreduction of carbon dioxide to methane in the presence of simulated sunshine irradiation. In more detail, the total yield of methane over the hierarchical flower-like structure of BiOI was measured at 0.198 μmol h^{-1} g^{-1}. In contrast, the total production of methane over bulk TiO_2 and BiOI alone was measured at 0.015 and 0.085 μmol h^{-1} g^{-1}, respectively. This study demonstrates that incorporating a 3D nanostructure into BiOI may effectively boost its photocatalytic capacity for carbon dioxide conversion (figure 3.8(b)). Hydro/solvothermal treatment is widely recognized as the preferred approach for fabricating three-dimensional structured BiOX due to its exceptional reaction efficiency and uncomplicated experimental methodology. The following three primary phases are often included in the creation mechanism of three-dimensional BiOX assemblies obtained via solvothermal and hydrothermal reactions. First, BiOX nuclei are produced; second, the development of two-dimensional NSs occurs through dissolution–renucleation. Lastly, the formation of three-dimensional BiOX structures takes place through the directed aggregation of two-dimensional NSs under the influence of an electric multipole field [81, 82]. In general, the greater PCA of three-dimensional nanostructures may be attributed to their higher capacity for curtate diffusion channels, light harvesting, more reactive areas, and quicker separation of photogenerated charge carriers.

3.9 Bismuth-rich strategy

Enhancing the PCA of BiOX materials is a viable alternative that can be achieved using the Bi-rich method. When compared to their unmodified forms, the BiOX materials enriched with Bi content (denoted as $Bi_aO_bX_c$) have been demonstrated to exhibit enhanced features such as improved optical properties, tunable electronic structures, and electrical conductivity. Compared to previous procedures, the Bi-enrichment strategy adds more Bi atoms to the final materials' surface. The high attraction that Bi has for N atoms contributes to an increase in the number of N_2 adsorption sites [83]. There are a number of ways (including calcination, hydrothermal procedures, and water-induced self-assembly) that have been proposed to enhance the amount of Bi that is included in BiOX photocatalysts [84, 85]. The water-induced self-assembly process takes much more time (about ten days), compared to the calcination method and the hydrothermal method [85]. The displacement response is yet another intriguing strategic approach. A simple displacement procedure was used to generate several other BiOX compounds enriched with Bi content ($Bi_{24}O_{31}Cl_{10}$, $Bi_{12}O_{17}Cl_2$, $Bi_{12}O_{15}C_{16}$, Bi_3O_4Cl, Bi_3O_4Br, $Bi_{24}O_{31}Br_{10}$, $Bi_4O_5Br_2$, Bi_5O_7I, $Bi_4O_5I_2$, and $Bi_7O_9I_3$) by subjecting Bi_2O_3 to an appropriately acidic environment, where the halogen species effectively displaced the oxygen atoms [86]. After 2 h of irradiation, the $Bi_7O_9I_3$ and $Bi_4O_5I_2$ that were produced demonstrated outstanding PCA for the breakdown of BPA under VL. the $Bi_7O_9I_3$ exhibited an elimination efficiency of 99.1%, whereas the $Bi_4O_5I_2$ had a slightly lower elimination efficiency of 93.0%. When compared to other BiOX materials, these particular ones had a substantially smaller band gap energy and a relatively better ability of oxidation (a more positive valance band potential) which are the reasons for their excellent performance. Calcination, which combines Bi_2O_3 and BiOX, is an example of another type of displacement process. Materials with the composition Bi_3O_4Cl were produced by calcining Bi_2O_3 and BiOCl at a temperature of 700 °C for 24 h in the presence of air [87]. By varying the molar ratios of BiOX to Bi_2O_3, it is possible to produce BiOX materials with various levels of the element Bi.

Calcination was the method that was used for the preparation of Bi_5O_7I in the presence of air [83, 88]. When heated to temperatures of more than 400 °C, the atoms of iodine were progressively changed into atoms of oxygen, producing the compound Bi_5O_7I [88]. The Bi_5O_7I exhibited significantly higher nitrogen fixation efficiency (81.73 μmol g^{-1} h^{-1}) than its pristine counterpart, BiOI, which had considerably poorer nitrogen fixation efficiency (21.96 μmol g^{-1} h^{-1}). This is most likely a result of the fact that the conduction band position of Bi_5O_7I is more negative in comparison to BiOI (−0.52 V as opposed to −0.12 V, respectively). According to the DFT calculations, the conduction band of the molecule $Bi_aO_bX_c$ is predominately made up of Bi 6p orbitals. In contrast, its VB is predominately made up of hybrids of O 2p and X np orbitals (n = 3, 4, and 5 when X = Cl, Br, and I, respectively) [89, 90]. Therefore, a shift in the component makeup of $Bi_aO_bX_c$ catalysts would result in a variation in band gap structure [91], and this shift would be related to the type of X atom. For instance, in the case of BiOBr and BiOCl, the band gap shrunk as the amount of Bi increased, which enhanced the substances' capacity to absorb VL. As a result, the valence band maximum (VBM) and the conduction band minimum (CBM) would move in opposite directions, with the CBM migrating downward and the VBM migrating upward. Because of this, the two of them working together may result in a band gap that is smaller compared to the pure BiOCl [83]. The optical response of the BiOCl enriched Bi content can encompass the expanded visible range as a result of its band gap being shorter, which is preferred for practical applications [92]. However, as a result of the downshift in the CBM location, the photoreduction potential of BiOCl is expected to diminish [92]. In contrast, a rise in the amount of Bi that is present in BiOI both downshifts VBM and upshifts CBM, which favorably improves BiOI's reduction potential. About the BiOBr catalyst, it was found that increasing the Bi content caused both CBM and VBM to increase, with the former upshift being more pronounced than the latter [91]. As a result, BiOBr's band gap decreased generally, increasing its capacity to absorb light. The band gap increases and Bi content increase in BiOI decreases its VL absorption. By varying the degree of dehalogenation, nonstoichiometric BiOX was created from the comparable stoichiometric BiOX. This was made possible because varying degrees of oxidation occurred at different annealing temperatures. In general, a greater annealing temperature led to more oxygen replacing the present halogens. The BiOI and Bi_5O_7I were produced by annealing a BiI_3 precursor at temperatures of 350 °C and 450 °C, respectively [93]. Furthermore, $Bi_{24}O_{31}Br_{10}$ specimens were synthesized by subjecting them to a range of heat-treatment temperatures spanning from 300 °C to 800 °C [94]. The $Bi_{24}O_{31}Br_{10}$ that was developed showed a high PCA for H_2 evolution when exposed to VL. It also had the greatest activity in the photocatalytic reduction of Cr (VI) out of the BiOBr, $Bi_{24}O_{31}Br_{10}$, and Bi_2O_3 that was tested.

The type of solvent that is used is an essential component in the synthesis of Bi-rich BiOX. Hierarchical MSs of $Bi_7O_9I_3$, $Bi_4O_5I_2$, and BiOI were produced via solvothermal processing at a temperature of 130 °C over 12 h. The types of solvents used were the sole variable [95]. Utilizing ethanol as the solvent allowed for the formation of BiOI, and utilizing glycerol and EG resulted in the production of $Bi_4O_5I_2$ and $Bi_7O_9I_3$, respectively. Finally, the preparation of Bi-rich BiOX may be done utilizing an open method combined with heat treatment [96–99].

An example of this would be the synthesis of the $Bi_7O_9I_3$ micro- and nanostructures at a temperature of 160 °C, utilizing an open three-neck round-bottom flask system with EG as the solvent [97]. In an open system operating at high temperatures, some iodide ions were removed from the solution, producing iodine ions, which eventually sublimed away. In the presence of VL, the hierarchical $Bi_7O_9I_3$ structure that was produced exhibited a greater PCA for the breakdown of phenol. After 4 h of irradiation, 94% of the phenol had been eliminated.

3.10 Carbonaceous materials compounding

Enhancing the photocatalytic efficacy of BiOX nanoparticles is significantly helped by the incorporation of a wide variety of carbonaceous materials, including carbon nanotubes (CNTs), biochar, carbon quantum dots (CQDs), and graphene, amongst others. Owing to its great mobility of electrons, huge specific surface area, and high conductivity [100–102], graphene is thought to be an effective electron collector and transporter in the process of photocatalysis. In light of these findings, RGO/BiOX nanocomposites display noticeably heightened PCA [103]. An *in situ* solvothermal method using graphene oxide (GO) was used by Li *et al* in the preparation of a BiOBr/graphene composite. After the GO was converted to graphene, surface BiOBr

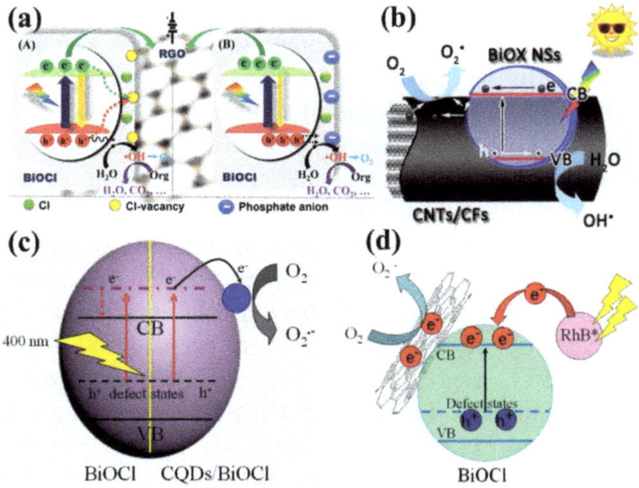

Figure 3.9. (a) Schematic pathway for the photoinduced charge separation and transfer in RGO/BOC photoanodes (A) before and (B) after modification of phosphate. (Reproduced with permission from [105]. Copyright 2017 Elsevier.) (b) The photocatalytic mechanism of the hierarchical CNT/CF-BiOX NS structures. (Reproduced with permission from [106]. Copyright 2014 Springer.) (c) Schematic of the transfer and separation of photoexcited charges in the CQD/BiOCl material combined with the possible reaction mechanism of the photocatalytic procedure. (Reproduced with permission from [108]. Copyright 2015 American Chemical Society.) (d) Proposed mechanism of degradation of RhB with MWCNT/BiOCl catalysts. (Reproduced with permission from [110]. Copyright 2016 John Wiley and Sons.)

NPs were produced. In terms of PCA, this combination is superior [104]. The graphene oxide was used by Zhang *et al* as the precursor material for synthesizing the reduced graphene oxide (RGO)/BiOCl hybrids [105]. The RGO/BiOCl photocatalyzed MO degradation 8.4 times and water oxidation 3.8 times better than BiOCl. This was attributed to the chemical bonding that occurred between BiOCl and RGO resulting in a longer carrier lifespan and speedier separation of charges. It has been demonstrated that empty chloride sites serve as recombination hubs that inhibit the separation of photocharge (figure 3.9(a)). Dong *et al* [103] created flower-like nanocomposites of BiOCl/RGO to break sulfanilamide (SN) when exposed to irradiation by natural sunshine. The sample that included 1 wt% RGO displayed good PCA owing to the enhanced VL absorption and improved electron transfer ability, rather than the changes in band gap and surface area value that were seen in the other samples. Xu *et al* coated hierarchical BiOX on the freshly manufactured functional CNTs using ionic layer reaction and adsorption. The exceptional mechanical capabilities and pollutant-absorbing capacity of nanotubes were used in tailoring the energy gaps of BiOX through the utilization of covalent bonds (figure 3.9(b)) [106]. Xiong *et al* employed the solvothermal method to fabricate composites of BiOI and multi-walled carbon nanotubes (MWCNT). These composites exhibited superior photocatalytic performance compared to pure BiOI, attributed to the enhanced interfacial charge transfer [107]. Recent developments in BiOX have resulted in incorporating CQDs, which are quasi-spherical nanoparticles that typically have a diameter of less than 10 nm. Di *et al* developed the BiOCl modified with CQDs, which demonstrated the improved PCA due to the higher electron transfer capacity of the CQDs (figure 3.9(c)) [108]. In addition, the N-doped CQD-modified BiOBr was developed by the same group to enhance the photoreduction of CIP, tetracycline hydrochloride, RhB, and BPA [109]. Using an ionic liquid-aided solvothermal technique, Yin *et al* [110] successfully produced MWCNT/BiOCl nanocomposites. The increased separation efficiency of photo-excited charge carriers was the main reason the as-obtained samples outperformed pure BiOCl in the photoreduction of RhB dye (figure 3.9(d)). However, the aforementioned carbonaceous NMs have a few drawbacks, such as a difficult production procedure and a prohibitively expensive price. As a result, biochar, a carbonaceous substance that is readily available and inexpensive, has recently come to the attention of many individuals [111, 112]. First, Li *et al* [113] made use of biochar to synthesize biochar/BiOX photocatalysts using a straightforward *in situ* precipitation technique. Biochar promoted BiOBr photocatalytic processes and BiOCl photosensitization reactions by acting as a benign electron carrier.

References

[1] Wang H, Zhang L, Chen Z, Hu J, Li S, Wang Z, Liu J and Wang X 2014 Semicondu ctor heterojunction photocatalysts: design, construction, and photocatalytic performances *Chem. Soc. Rev.* **43** 5234–44

[2] Ning S, Ding L, Lin Z, Lin Q, Zhang H, Lin H, Long J and Wang X 2016 One-pot fabrication of Bi_3O_4Cl/BiOCl plate-on-plate heterojunction with enhanced visible-light photocatalytic activity *Appl. Catal.* B **185** 203–12

[3] Peng Y, Yu P P, Chen Q G, Zhou H Y and Xu A W 2015 Facile fabrication of $Bi_{12}O_{17}Br_2$/$Bi_24O_{31}Br_{10}$ type II heterostructures with high visible photocatalytic activity *J. Phys. Chem.* C **119** 13032–40

[4] Duo F, Wang Y, Fan C, Zhang X and Wang Y 2016 Enhanced visible light photocatalytic activity and stability of CQDs/BiOBr composites: the upconversion effect of CQDs *J. Alloys Compd.* **685** 34–41

[5] Xia J, Di J, Li H, Xu H, Li H and Guo S 2016 Ionic liquid-induced strategy for carbon quantum dots/BiOX (X = Br, Cl) hybrid nanosheets with superior visible light-driven photocatalysis *Appl. Catal.* B **181** 260–9

[6] Gao F, Zeng D, Huang Q, Tian S and Xie C 2012 Chemically bonded graphene/BiOCl nanocomposites as high-performance photocatalysts *Phys. Chem. Chem. Phys.* **14** 10572–8

[7] Liu H, Su Y, Chen Z, Jin Z and Wang Y 2014 Graphene sheets grafted three-dimensional $BiOBr_{0.2}I_{0.8}$ microspheres with excellent photocatalytic activity under visible light *J. Hazard. Mater.* **266** 75–83

[8] Di J, Xia J, Yin S, Xu H, Xu L, Xu Y, He M and Li H 2014 Preparation of sphere-like g-C_3N_4/BiOI photocatalysts via a reactable ionic liquid for visible-light-driven photocatalytic degradation of pollutants *J. Mater. Chem.* A **2** 5340–51

[9] Ikram M, Bari M A and Haider J 2023 *Photocatalytic Dye Degradation Using Green Polymeric-based Nanostructures: Principles and Applications in Wastewater Treatment* (Bristol: Institute of Physics Publishing)

[10] Li X, Wang T, Tao X, Qiu G, Li C and Li B 2020 Interfacial synergy of Pd sites and defective BiOBr for promoting the solar-driven selective oxidation of toluene *J. Mater. Chem.* A **8** 17657–69
[11] Yu C, Cao F, Li G, Wei R, Yu J C, Jin R, Fan Q and Wang C 2013 Novel noble metal (Rh, Pd, Pt)/BiOX(Cl, Br, I) composite photocatalysts with enhanced photocatalytic performance in dye degradation *Sep. Purif. Technol.* **120** 110–22
[12] Arumugam M, Koutavarapu R, Seralathan K K, Praserthdam S and Praserthdam P 2023 Noble metals (Pd, Ag, Pt, and Au) doped bismuth oxybromide photocatalysts for improved visible light-driven catalytic activity for the degradation of phenol *Chemosphere* **324** 138368
[13] Wang W, Dai R, Zhang L, Wu Q, Wang X, Zhang S, Shao T, Zhang F, Yan J and Zhang W 2020 Experimental and DFT investigation on the different effects of Er^{3+}- and Ag^{+}-doped BiOBr microspheres in enhancing photocatalytic activity under visible light irradiation *J. Mater. Sci.* **55** 11226–40
[14] Yang W, Wen Y, Chen R, Zeng D and Shan B 2014 Study of structural, electronic and optical properties of tungsten doped bismuth oxychloride by DFT calculations *Phys. Chem. Chem. Phys.* **16** 21349–55
[15] Xia J, Xu L, Zhang J, Yin S, Li H, Xu H and Di J 2013 Improved visible light photocatalytic properties of Fe/BiOCl microspheres synthesized via self-doped reactable ionic liquids *CrystEngComm.* **15** 10132–41
[16] Jiang G, Wang R, Wang X, Xi X, Hu R, Zhou Y, Wang S, Wang T and Chen W 2012 Novel highly active visible-light-induced photocatalysts based on BiOBr with Ti doping and Ag decorating *ACS Appl. Mater. Interfaces* **4** 4440–4
[17] Chen X, Zhang X, Li Y H, Qi M Y, Li J Y, Tang Z R, Zhou Z and Xu Y J 2021 Transition metal doping BiOBr nanosheets with oxygen vacancy and exposed {102} facets for visible light nitrogen fixation *Appl. Catal.* B **281** 119516
[18] Chen X, Liu L and Huang F 2015 Black titanium dioxide (TiO_2) nanomaterials *Chem. Soc. Rev.* **44** 1861–85
[19] Yu C, He H, Fan Q, Xie W, Liu Z and Ji H 2019 Novel B-doped BiOCl nanosheets with exposed (001) facets and photocatalytic mechanism of enhanced degradation efficiency for organic pollutants *Sci. Total Environ.* **694** 133727
[20] Wu D, Yue S, Wang W, An T, Li G, Yip H Y, Zhao H and Wong P K 2016 Boron-doped BiOBr nanosheets with enhanced photocatalytic inactivation of *Escherichia coli Appl. Catal.* B **192** 35–45
[21] Liu Z S, Liu J L, Wang H Y, Cao G and Niu J N 2016 Boron-doped bismuth oxybromide microspheres with enhanced surface hydroxyl groups: synthesis, characterization and dramatic photocatalytic activity *J. Colloid Interface Sci.* **463** 324–31
[22] Obeid M M *et al* 2020 First-principles investigation of nonmetal doped single-layer BiOBr as a potential photocatalyst with a low recombination rate *Phys. Chem. Chem. Phys.* **22** 15354–64
[23] Zeng L, Zhe F, Wang Y, Zhang Q, Zhao X, Hu X, Wu Y and He Y 2019 Preparation of interstitial carbon doped BiOI for enhanced performance in photocatalytic nitrogen fixation and methyl orange degradation *J. Colloid Interface Sci.* **539** 563–74
[24] Qu S, Xiong Y and Zhang J 2018 Graphene oxide and carbon nanodots co-modified BiOBr nanocomposites with enhanced photocatalytic 4-chlorophenol degradation and mechanism insight *J. Colloid Interface Sci.* **527** 78–86

[25] Jiang G, Li X, Wei Z, Jiang T, Du X and Chen W 2014 Growth of N-doped BiOBr nanosheets on carbon fibers for photocatalytic degradation of organic pollutants under visible light irradiation *Powder Technol.* **260** 84–9

[26] Jiang Z, Liu Y, Jing T, Huang B, Wang Z, Zhang X, Qin X and Dai Y 2015 One-pot solvothermal synthesis of S-doped BiOCl for solar water oxidation *RSC Adv.* **5** 47261–4

[27] Wang C Y, Zeng Q and Zhu G 2021 Novel S-doped BiOBr nanosheets for the enhanced photocatalytic degradation of bisphenol A under visible light irradiation *Chemosphere.* **268** 128854

[28] Zhang L, Liu F, Xiao X, Zuo X and Nan J 2019 Microwave synthesis of iodine-doped bismuth oxychloride microspheres for the visible light photocatalytic removal of toxic hydroxyl-contained intermediates of parabens: catalyst synthesis, characterization, and mechanism insight *Environ. Sci. Pollut. Res.* **26** 28871–83

[29] He M, Li W, Xia J, Xu L, Di J, Xu H, Yin S, Li H and Li M 2015 The enhanced visible light photocatalytic activity of yttrium-doped BiOBr synthesized via a reactable ionic liquid *Appl. Surf. Sci.* **331** 170–8

[30] Zhong S, Wang X, Wang Y, Zhou F, Li J, Liang S and Li C 2020 Preparation of Y^{3+}-doped BiOCl photocatalyst and its enhancing effect on degradation of tetracycline hydrochloride wastewater *J. Alloys Compd.* **843** 155598

[31] Xia J, Ji M, Li W, Di J, Xu H, He M, Zhang Q and Li H 2016 Synthesis of erbium ions doped BiOBr via a reactive ionic liquid with improved photocatalytic activity *Colloids Surf.* A **489** 343–50

[32] Dash A, Sarkar S, Adusumalli V N K B and Mahalingam V 2014 Microwave synthesis, photoluminescence, and photocatalytic activity of PVA-functionalized Eu^{3+}-doped BiOX (X = Cl, Br, I) nanoflakes *Langmuir* **30** 1401–9

[33] Yin S, Fan W, Di J, Wu T, Yan J, He M, Xia J and Li H 2017 La^{3+}-doped BiOBr microsphere with enhanced visible light photocatalytic activity *Colloids Surf.* A **513** 160–7

[34] Li H, Shang J, Ai Z and Zhang L 2015 Efficient visible light nitrogen fixation with BiOBr nanosheets of oxygen vacancies on the exposed {001} facets *J. Am. Chem. Soc.* **137** 6393–9

[35] Luo S, Xu J, Li Z, Liu C, Chen J, Min X, Fang M and Huang Z 2017 Bismuth oxyiodide coupled with bismuth nanodots for enhanced photocatalytic bisphenol A degradation: synergistic effects and mechanistic insight *Nanoscale.* **9** 15484–93

[36] Wang H, Yong D, Chen S, Jiang S, Zhang X, Shao W, Zhang Q, Yan W, Pan B and Xie Y 2018 Oxygen-vacancy-mediated exciton dissociation in BiOBr for boosting charge-carrier-involved molecular oxygen activation *J. Am. Chem. Soc.* **140** 1760–6

[37] Li H, Qin F, Yang Z, Cui X, Wang J and Zhang L 2017 New reaction pathway induced by plasmon for selective benzyl alcohol oxidation on BiOCl possessing oxygen vacancies *J. Am. Chem. Soc.* **139** 3513–21

[38] Wu S, Xiong J, Sun J, Hood Z D, Zeng W, Yang Z, Gu L, Zhang X and Yang S Z 2017 Hydroxyl-dependent evolution of oxygen vacancies enables the regeneration of BiOCl photocatalyst *ACS Appl. Mater. Interfaces* **9** 16620–6

[39] Ye L, Deng K, Xu F, Tian L, Peng T and Zan L 2012 Increasing visible-light absorption for photocatalysis with black BiOCl *Phys. Chem. Chem. Phys.* **14** 82–5

[40] Wang X J, Zhao Y, Li F T, Dou L J, Li Y P, Zhao J and Hao Y J 2016 A chelation strategy for *in situ* constructing surface oxygen vacancy on {001} facets exposed BiOBr nanosheets *Sci. Rep.* **6** 24918

[41] Di J *et al* 2018 Defect-rich $Bi_{12}O_{17}Cl_2$ nanotubes self-accelerating charge separation for boosting photocatalytic CO_2 reduction *Angew. Chem., Int. Ed.* **57** 14847–51
[42] Di J *et al* 2019 Defect-tailoring mediated electron–hole separation in single-unit-cell Bi_3O_4Br nanosheets for boosting photocatalytic hydrogen evolution and nitrogen fixation *Adv. Mater.* **31** 1807576
[43] Xue X *et al* 2018 Oxygen vacancy engineering promoted photocatalytic ammonia synthesis on ultrathin two-dimensional bismuth oxybromide nanosheets *Nano Lett.* **18** 7372–7
[44] Huang Y, Li H, Balogun M S, Liu W, Tong Y, Lu X and Ji H 2014 Oxygen vacancy induced bismuth oxyiodide with remarkably increased visible-light absorption and superior photocatalytic performance *ACS Appl. Mater. Interfaces* **6** 22920–7
[45] Wang Q, Liu Z, Liu D, Wang W, Zhao Z, Cui F and Li G 2019 Oxygen vacancy-rich ultrathin sulfur-doped bismuth oxybromide nanosheet as a highly efficient visible-light responsive photocatalyst for environmental remediation *Chem. Eng. J.* **360** 838–47
[46] Cui D, Wang L, Xu K, Ren L, Weng L, Yu Y, Du Y and Hao W 2018 Band-gap engineering of BiOCl with oxygen vacancies for efficient photooxidation properties under visible-light irradiation *J. Mater. Chem.* A **6** 2193–9
[47] Chen M, Yu S, Zhang X, Wang F, Lin Y and Zhou Y 2016 Insights into the photo-sensitivity of BiOCl nanoplates with exposing {001} facets: the role of oxygen vacancy *Superlattices Microstruct.* **89** 275–81
[48] Di J, Yan C, Handoko A D, Seh Z W, Li H and Liu Z 2018 Ultrathin two-dimensional materials for photo- and electrocatalytic hydrogen evolution *Mater. Today* **21** 749–70
[49] Xiong J, Di J and Li H 2018 Atomically thin 2D multinary nanosheets for energy-related photo, electrocatalysis *Adv. Sci.* **5** 1800244
[50] Li J, Zhang L, Li Y and Yu Y 2014 Synthesis and internal electric field dependent photoreactivity of Bi_3O_4Cl single-crystalline nanosheets with high {001} facet exposure percentages *Nanoscale.* **6** 167–71
[51] Wang C Y, Zhang X, Qiu H B, Huang G X and Yu H Q 2017 $Bi_{24}O_{31}Br_{10}$ nanosheets with controllable thickness for visible-light-driven catalytic degradation of tetracycline hydrochloride *Appl. Catal.* B **205** 615–23
[52] Feng J, Qian X, Huang C W and Li J 2012 Strain-engineered artificial atom as a broad-spectrum solar energy funnel *Nat. Photonics* **6** 866–72
[53] Feng H *et al* 2015 Modulation of photocatalytic properties by strain in 2D BiOBr nanosheets *ACS Appl. Mater. Interfaces* **7** 27592–6
[54] Wang L, Wang L, Du Y, Xu X and Dou S X 2021 Progress and perspectives of bismuth oxyhalides in catalytic applications *Mater. Today Phys.* **16** 100294
[55] Shiri D, Kong Y, Buin A and Anantram M P 2008 Strain induced change of bandgap and effective mass in silicon nanowires *Appl. Phys. Lett.* **93** 073114
[56] Yang H G, Sun C H, Qiao S Z, Zou J, Liu G, Smith S C, Cheng H M and Lu G Q 2008 Anatase TiO_2 single crystals with a large percentage of reactive facets *Nature* **453** 638–41
[57] Jiang Z Y, Kuang Q, Xie Z X and Zheng L S 2010 Syntheses and properties of micro/nanostructured crystallites with high-energy surfaces *Adv. Funct. Mater.* **20** 3634–45
[58] Martin D J, Liu G, Moniz S J A, Bi Y, Beale A M, Ye J and Tang J 2015 Efficient visible driven photocatalyst, silver phosphate: performance, understanding and perspective *Chem. Soc. Rev.* **44** 7808–28
[59] Liu G, Yang H G, Pan J, Yang Y Q, Lu G Q M and Cheng H M 2014 Titanium dioxide crystals with tailored facets *Chem. Rev.* **114** 9559–612

[60] Peng S, Li L, Zhu P, Wu Y, Srinivasan M, Mhaisalkar S G, Ramakrishna S and Yan Q 2013 Controlled synthesis of BiOCl hierarchical self-assemblies with highly efficient photocatalytic properties *Chem. Asian J.* **8** 258–68

[61] Jiang J, Zhao K, Xiao X and Zhang L 2012 Synthesis and facet-dependent photoreactivity of BiOCl single-crystalline nanosheets *J. Am. Chem. Soc.* **134** 4473–6

[62] Zhang H, Yang Y, Zhou Z, Zhao Y and Liu L 2014 Enhanced photocatalytic properties in BiOBr nanosheets with dominantly exposed (102) facets *J. Phys. Chem.* C **118** 14662–9

[63] Pan M, Zhang H, Gao G, Liu L and Chen W 2015 Facet-dependent catalytic activity of nanosheet-assembled bismuth oxyiodide microspheres in degradation of bisphenol A *Environ. Sci. Technol.* **49** 6240–8

[64] Ong W J, Tan L L, Chai S P, Yong S T and Mohamed A R 2014 Facet-dependent photocatalytic properties of TiO_2-based composites for energy conversion and environmental remediation *ChemSusChem.* **7** 690–719

[65] Wu X, Ng Y H, Wang L, Du Y, Dou S X, Amal R and Scott J 2017 Improving the photo-oxidative capability of BiOBr: via crystal facet engineering *J. Mater. Chem.* A **5** 8117–24

[66] Zhang X, Wang L W, Wang C Y, Wang W K, Chen Y L, Huang Y X, Li W W, Feng Y J and Yu H Q 2015 Synthesis of $BiOCl_xBr_{1-x}$ nanoplate solid solutions as a robust photocatalyst with tunable band structure *Chem. Eur. J.* **21** 11872–7

[67] Zhang G, Cai L, Zhang Y and Wei Y 2018 Bi^{5+}, $Bi^{(3-x)+}$, and oxygen vacancy induced $BiOCl_xI_{1-x}$ solid solution toward promoting visible-light driven photocatalytic activity *Chem. Eur. J.* **24** 7434–44

[68] Zhang G, Zhang L, Liu Y, Liu L, Huang C P, Liu H and Li J 2016 Substitution boosts charge separation for high solar-driven photocatalytic performance *ACS Appl. Mater. Interfaces* **8** 26783–93

[69] Yu Y, Xu M and Dong S 2016 Photoenergy storage and power amplification strategy in membrane-less photoelectrochemical biofuel cells *Chem. Commun.* **52** 6716–9

[70] Liu W, Shang Y, Zhu A, Tan P, Liu Y, Qiao L, Chu D, Xiong X and Pan J 2017 Enhanced performance of doped BiOCl nanoplates for photocatalysis: understanding from doping insight into improved spatial carrier separation *J. Mater. Chem.* A **5** 12542–9

[71] Huang Y, Long B, Li H, Balogun M S, Rui Z, Tong Y and Ji H 2015 Enhancing the photocatalytic performance of $BiOCl_xI_{1-x}$ by introducing surface disorders and bi nanoparticles as cocatalyst *Adv. Mater. Interfaces* **2** 1500249

[72] Liu G, Wang T, Ouyang S, Liu L, Jiang H, Yu Q, Kako T and Ye J 2015 Band-structure-controlled $BiO(ClBr)_{(1-x)/2}I_x$ solid solutions for visible-light photocatalysis *J. Mater. Chem.* A **3** 8123–32

[73] Jia X, Cao J, Lin H, Zhang M, Guo X and Chen S 2017 Transforming type-I to type-II heterostructure photocatalyst via energy band engineering: a case study of I-BiOCl/I-BiOBr *Appl. Catal.* B **204** 505–14

[74] Gnayem H and Sasson Y 2013 Hierarchical nanostructured 3D flowerlike $BiOCl_xBr_{1-x}$ semiconductors with exceptional visible light photocatalytic activity *ACS Catal.* **3** 186–91

[75] Zhang X, Wang C Y, Wang L W, Huang G X, Wang W K and Yu H Q 2016 Fabrication of $BiOBr_xI_{1-x}$ photocatalysts with tunable visible light catalytic activity by modulating band structures *Sci. Rep.* **6** 22800

[76] Liu Y, Son W J, Lu J, Huang B, Dai Y and Whangbo M H 2011 Composition dependence of the photocatalytic activities of $BiOCl_{1-x}Br_x$ solid solutions under visible light *Chem. Eur. J.* **17** 9342–9

[77] Xu P *et al* 2012 Use of iron oxide nanomaterials in wastewater treatment: a review *Sci. Total Environ.* **424** 1–10

[78] Huang D, Wang Y, Zhang C, Zeng G, Lai C, Wan J, Qin L and Zeng Y 2016 Influence of morphological and chemical features of biochar on hydrogen peroxide activation: implications on sulfamethazine degradation *RSC Adv.* **6** 73186–96

[79] Zhang G, Su A, Qu J and Xu Y 2014 Synthesis of BiOI flowerlike hierarchical structures toward photocatalytic reduction of CO_2 to CH_4 *Mater. Res. Bull.* **55** 43–7

[80] Zhao Y, Tan X, Yu T and Wang S 2015 SDS-assisted solvothermal synthesis of BiOBr microspheres with highly visible-light photocatalytic activity *Mater. Lett.* **164** 243–7

[81] Huo Y, Zhang J, Miao M and Jin Y 2012 Solvothermal synthesis of flower-like BiOBr microspheres with highly visible-light photocatalytic performances *Appl. Catal. B: Environ.* B **111–2** 334–41

[82] Xia J, Yin S, Li H, Xu H, Yan Y and Zhang Q 2011 Self-assembly and enhanced photocatalytic properties of BiOI hollow microspheres via a reactable ionic liquid *Langmuir* **27** 1200–6

[83] Xiong J, Song P, Di J and Li H 2020 Bismuth-rich bismuth oxyhalides: a new opportunity to trigger high-efficiency photocatalysis *J. Mater. Chem.* A **8** 21434–54

[84] Gao K, Zhang C, Zhang Y, Zhou X, Gu S, Zhang K, Wang X and Song X 2022 Oxygen vacancy engineering of novel ultrathin $Bi_{12}O_{17}Br_2$ nanosheets for boosting photocatalytic N_2 reduction *J. Colloid Interface Sci.* **614** 12–23

[85] Li P *et al* 2020 Visible-light-driven nitrogen fixation catalyzed by Bi_5O_7Br nanostructures: enhanced performance by oxygen vacancies *J. Am. Chem. Soc.* **142** 12430–9

[86] Xiao X, Liu C, Hu R, Zuo X, Nan J, Li L and Wang L 2012 Oxygen-rich bismuth oxyhalides: generalized one-pot synthesis, band structures and visible-light photocatalytic properties *J. Mater. Chem.* **22** 22840–3

[87] Lin X, Huang T, Huang F, Wang W and Shi J 2006 Photocatalytic activity of a Bi-based oxychloride Bi_3O_4Cl *J. Phys. Chem.* B **110** 24629–34

[88] Lan M, Zheng N, Dong X, Hua C, Ma H and Zhang X 2020 Bismuth-rich bismuth oxyiodide microspheres with abundant oxygen vacancies as an efficient photocatalyst for nitrogen fixation *Dalton Trans.* **49** 9123–9

[89] Di J, Xia J, Li H, Guo S and Dai S 2017 Bismuth oxyhalide layered materials for energy and environmental applications *Nano Energy.* **41** 172–92

[90] Bai Y, Ye L, Chen T, Wang P, Wang L, Shi X and Wong P K 2017 Synthesis of hierarchical bismuth-rich $Bi_4O_5Br_xI_{2-x}$ solid solutions for enhanced photocatalytic activities of CO_2 conversion and Cr(VI) reduction under visible light *Appl. Catal.* B **203** 633–40

[91] Li J, Li H, Zhan G and Zhang L 2017 Solar water splitting and nitrogen fixation with layered bismuth oxyhalides *Acc. Chem. Res.* **50** 112–21

[92] Cui P, Wang J, Wang Z, Chen J, Xing X, Wang L and Yu R 2016 Bismuth oxychloride hollow microspheres with high visible light photocatalytic activity *Nano Res.* **9** 593–601

[93] Ai L, Zeng Y and Jiang J 2014 Hierarchical porous BiOI architectures: facile microwave nonaqueous synthesis, characterization and application in the removal of Congo red from aqueous solution *Chem. Eng. J.* **235** 331–9

[94] Shang J, Hao W, Lv X, Wang T, Wang X, Du Y, Dou S, Xie T, Wang D and Wang J 2014 Bismuth oxybromide with reasonable photocatalytic reduction activity under visible light *ACS Catal.* **4** 954–61

[95] Liu Q C, Ma D K, Hu Y Y, Zeng Y W and Huang S M 2013 Various bismuth oxyiodide hierarchical architectures: alcohothermal-controlled synthesis, photocatalytic activities, and adsorption capabilities for phosphate in water *ACS Appl. Mater. Interfaces* **5** 11927–34

[96] Xiao X, Hao R, Zuo X, Nan J, Li L and Zhang W 2012 Microwave-assisted synthesis of hierarchical $Bi_7O_9I_3$ microsheets for efficient photocatalytic degradation of bisphenol-A under visible light irradiation *Chem. Eng. J.* **209** 293–300

[97] Xiao X and Zhang W D 2011 Hierarchical $Bi_7O_9I_3$ micro/nano-architecture: facile synthesis, growth mechanism, and high visible light photocatalytic performance *RSC Adv.* **1** 1099–105

[98] Liu H, Su Y, Chen Z, Jin Z and Wang Y 2014 $Bi_7O_9I_3$/reduced graphene oxide composite as an efficient visible-light-driven photocatalyst for degradation of organic contaminants *J. Mol. Catal.* A **391** 175–82

[99] Su Y, Wang H, Ye L, Jin X, Xie H, He C and Bao K 2014 Shape-dependent photocatalytic activity of Bi_5O_7I caused by facets synergetic and internal electric field effects *RSC Adv.* **4** 65056–64

[100] Katsnelson M I 2012 *Graphene: Carbon in Two Dimensions* (Cambridge: Cambridge University Press)

[101] Wang H, Yuan X, Wu Y, Huang H, Peng X, Zeng G, Zhong H, Liang J and Ren M M 2013 Graphene-based materials: fabrication, characterization and application for the decontamination of wastewater and wastegas and hydrogen storage/generation *Adv. Colloid Interface Sci.* 195–6 19–40

[102] Wang H, Yuan X, Zeng G, Wu Y, Liu Y, Jiang Q and Gu S 2015 Three-dimensional graphene based materials: synthesis and applications from energy storage and conversion to electrochemical sensor and environmental remediation *Adv. Colloid Interface Sci.* **221** 41–59

[103] Dong S, Pi Y, Li Q, Hu L, Li Y, Han X, Wang J and Sun J 2016 Solar photocatalytic degradation of sulfanilamide by BiOCl/reduced graphene oxide nanocomposites: mechanism and degradation pathways *J. Alloys Compd.* **663** 1–9

[104] Tu X, Luo S, Chen G and Li J 2012 One-pot synthesis, characterization, and enhanced photocatalytic activity of a biobr-graphene composite *Chem. Eur. J.* **18** 14359–66

[105] Li Z, Qu Y, Hu K, Humayun M, Chen S and Jing L 2017 Improved photoelectrocatalytic activities of BiOCl with high stability for water oxidation and MO degradation by coupling RGO and modifying phosphate groups to prolong carrier lifetime *Appl. Catal.* B **203** 355–62

[106] Weng B, Xu F and Xu J 2014 Hierarchical structures constructed by BiOX (X = Cl, I) nanosheets on CNTs/carbon composite fibers for improved photocatalytic degradation of methyl orange *J. Nanopart. Res.* **16** 2766

[107] Su M, He C, Zhu L, Sun Z, Shan C, Zhang Q, Shu D, Qiu R and Xiong Y 2012 Enhanced adsorption and photocatalytic activity of BiOI–MWCNT composites towards organic pollutants in aqueous solution *J. Hazard. Mater.* **229–30** 72–82

[108] Di J, Xia J, Ji M, Wang B, Yin S, Zhang Q, Chen Z and Li H 2015 Carbon quantum dots modified BiOCl ultrathin nanosheets with enhanced molecular oxygen activation ability for broad spectrum photocatalytic properties and mechanism insight *ACS Appl. Mater. Interfaces* **7** 20111–23

[109] Di J, Xia J, Ji M, Wang B, Li X, Zhang Q, Chen Z and Li H 2016 Nitrogen-doped carbon quantum dots/BiOBr ultrathin nanosheets: *in situ* strong coupling and improved molecular oxygen activation ability under visible light irradiation *ACS Sustain. Chem. Eng.* **4** 136–46

[110] Yin S, Di J, Li M, Fan W, Xia J, Xu H, Sun Y and Li H 2016 Synthesis of multiwalled carbon nanotube modified BiOCl microspheres with enhanced visible-light response photoactivity *Clean—Soil, Air, Water* **44** 781–7

[111] Zhang C, Lai C, Zeng G, Huang D, Yang C, Wang Y, Zhou Y and Cheng M 2016 Efficacy of carbonaceous nanocomposites for sorbing ionizable antibiotic sulfamethazine from aqueous solution *Water Res.* **95** 103–12

[112] Huang D, Liu L, Zeng G, Xu P, Huang C, Deng L, Wang R and Wan J 2017 The effects of rice straw biochar on indigenous microbial community and enzymes activity in heavy metal-contaminated sediment *Chemosphere.* **174** 545–53

[113] Li M, Huang H, Yu S, Tian N, Dong F, Du X and Zhang Y 2016 Simultaneously promoting charge separation and photoabsorption of BiOX (X = Cl, Br) for efficient visible-light photocatalysis and photosensitization by compositing low-cost biochar *Appl. Surf. Sci.* **386** 285–95

Chapter 4

Air purification and energy applications

Photocatalysis has garnered significant attention in the realm of solar energy conversion, as it provides an effective avenue for the advancement of sustainable and renewable sources of energy. Consequently, this technology holds the potential to address the pressing energy and environmental challenges that society currently faces. The aforementioned uses encompass the processes of water splitting to produce oxygen and hydrogen, conversion of carbon dioxide into hydrocarbon fuels, and photocatalytic reduction of nitrogen. Over recent decades, BiOX materials have emerged as a prevalent choice for photocatalysis due to their cost-effectiveness, lack of toxicity, and notable efficacy when exposed to light. This chapter presents a comprehensive examination of the oxygen evolution, hydrogen production, nitrogen reduction, and carbon dioxide reduction processes facilitated by BiOX materials. Furthermore, there is a discussion regarding organic synthesis carried out using BiOX materials. A comprehensive analysis of the mechanisms underlying each application of BiOX is presented.

4.1 Photocatalyst

In the process of water splitting, photocatalytic materials are typically made up of d^0 or d^{10} metal oxides, sulfides, nitrides, etc. However, there has been a recent shift in focus toward metal hetero-anionic materials, including metal oxynitrides, oxysulfides, oxyhalides, etc [1–3]. Within the scope of this category of substances, we will be concentrating our attention on bismuth oxyhalides (BiOX). The photocatalytic breakdown of pollutants, water splitting for hydrogen (H_2) generation, organic synthesis, photoreduction of carbon dioxide (CO_2) into fuels, molecular oxygen activation, and nitrogen (N_2) fixation are a few examples of the photocatalytic activities that can be evaluated using BiOX.

doi:10.1088/978-0-7503-5934-4ch4

4.2 Photocatalytic mechanisms

A constant redox reaction occurs during photocatalysis between the surface adsorbate and the catalyst [4]. In the course of the procedure, a photocatalyst is exposed to light, which, for optimal efficacy, necessitates an excitation energy source that surpasses or equals the band gap of the material ($h\nu \geqslant E_g$). Photoinduced charge carriers encompass both holes (h^+) located in the valence band with oxidation activity, as well as electrons (e^-) situated in the conduction band with reduction activity created in the energy band [5, 6]. Consumption of photogenerated carriers may be broken down into two distinct categories. First, most carriers generated by photoexcitation undergo recombination and then release the surplus energy by mechanisms that do not involve radiation. This might cause a decrease in the photocatalyst's effectiveness. Second, they go toward the photocatalyst's surface as well as react with the material's surface adsorbed there, such as oxygen, water, and hydroxyl radicals, to produce a different active species, including superoxide anion ($^{\bullet}O_2^-$) radicals, hydroxyl ($^{\bullet}OH$) radicals [7, 8].

The valance band of semiconductors based on bismuth (Bi) is made up of hybrid orbitals that combine Bi 6s and O 2p. Bi is classified as a fifth group member in the sixth period of the periodic table of elements. In its most common form, Bi exists as the Bi^{3+} ion, with an electronic configuration of $6s^2\ 6p^3$. The overlap between the O 1s as well as Bi 6s orbitals in the valence band is facilitated by the distortion caused by the presence of a lone pair in the Bi 6s orbital. This overlap results in a lowering of the band gap and an increase in the mobility of photoinduced charge carriers. When there is no electron in the 6s orbital, the valence state of Bi^{5+} has the potential to absorb visible light (VL) very well [9]. Band gaps of fewer than 3 eV are necessary for the activity of BiOX photocatalysts. It is worth noting that the valence band of all BiOX photocatalysts exhibits a higher positive potential compared to the redox potential of $^{\bullet}OH/OH^-$ (+1.99 eV), which suggests that the generation of $^{\bullet}OH$ can occur.

4.3 Photocatalytic oxygen (O_2) evolution

The methodology of water splitting by photocatalysis is widely recognized as a potential way to tackle the problems associated with energy and ecology. Due to the complicated nature of the four-electron redox mechanism, the reaction of O_2 evolution often slows down the entire water splitting process the most. The process of O_2 evolution is a redox reaction in which holes participate that necessitates the valence band edge to be positioned at a greater positive potential compared to oxidation half-reaction. In a recent study, the utilization of BiOCl materials for photocatalytic O_2 evolution was successfully reported. This process involved subjecting the materials to full-spectrum irradiation using a 300 W xenon lamp, while sodium iodate was employed as the electron-sacrificing agent [10]. The evolution efficacy of O_2 from water was significantly enhanced in the presence of whole spectrum irradiation due to the combined exploitation of Schottky junctions on (110) facets for carrier trapping as well as the hot hole injection of plasmonic silver nanocrystals on (001) facets. Furthermore, the incorporation of elemental

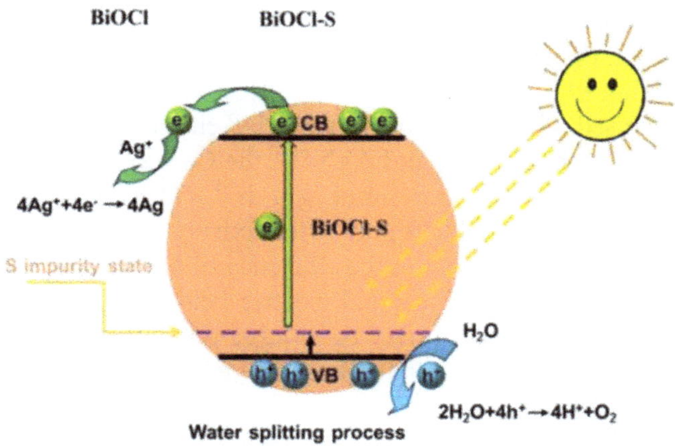

Figure 4.1. Photocatalytic mechanism of BiOCl-S. (Adapted with permission from [11]. Copyright 2015 The Royal Society of Chemistry.)

sulfur (S) into BiOCl was performed to greater the photocatalytic activity (PCA) for O_2 evolution. This process was carried out by subjecting the material to UV and VL light, with silver nitrate as the sacrificial reagent [11]. The introduction of S doping resulted in a notable enhancement in the efficacy of separating the photoinduced pairs of electron and hole. This improvement, therefore, led to a much higher rate of O_2 evolution, approximately five times higher in comparison observed in the absence of S doping in BiOCl (figure 4.1). Furthermore, to enhance the internal electric field (IEF) strength and obtain a separation of charge efficacy of up to 80%, homogeneous carbon (C)-doped Bi_3O_4Cl NSs were synthesized alongside BiOCl [12]. The hole and electron pairs were successfully separated by the strong IEF, as was confirmed by femtosecond-resolved transient absorption spectroscopy. Furthermore, the movement of hole and electron pairs from the interior of the material to its surface was found to be confined inside the [Cl] and [Bi_3O_4] layers, respectively. The C-doped Bi_3O_4Cl NSs exhibited the capability to split water and generate oxygen efficiently. This process achieved an efficiency of approximately 90 μmol l^{-1} under VL conditions (using a 150 W xenon arc lamp equipped with a 420 nm cutoff filter), without requiring the presence of an electron-scavenger or a co-catalyst. Because the conduction band position is much positive in comparison to H_2 production potential, the C-doped Bi_3O_4Cl NSs unfortunately cannot produce H_2.

The rate-determining phase in the procedure of water splitting, which involves the creation of O_2 and H_2, is commonly ascribed to the photocatalytic generation of O_2. In contrast to the two-electron transfer involved in the conversion of H^+ to H_2, the production of O_2 necessitates the involvement of four electron holes to facilitate the oxidation of H_2O and OH^-, enabling the acquisition of O_2 and overcoming a substantial kinetic obstacle. Therefore, it is necessary to enhance the oxidation half-reaction of water splitting to achieve a greater level of efficiency in H_2 production. BiOCl and BiOBr have both been the subject of research in the past, with the results of such investigations having been publicized [10, 13]. It is possible to reach a higher level of activity for the synthesis of oxygen using the $Bi_xO_yX_z$ materials that are already in use. Ning *et al* [14] created a heterostructure consisting of Bi_3O_4Cl and BiOCl and then used it to generate oxygen by employing a Z-scheme. Under electron scavengers such as $FeCl_3$ and $AgNO_3$, the O_2 production rate by Bi_3O_4Cl/BiOCl was measured to be 58.6 μmol g^{-1} h^{-1}. The utilization of ultrathin heterostructure systems has been found to effectively facilitate electron transfer, leading to a sufficient improvement in the efficiency of photocatalysis processes. The ultrathin heterojunction with an internal electric field (IEF) exhibits a significant charge transfer due to the appropriate band edge potentials and strong electronic interaction between two-dimensional Bi_3O_4Cl and BiOCl. This conclusion is supported by photoluminescence, theoretical computations, and photoelectricity tests. The improved photoefficiency of the ultrathin Bi_3O_4Cl/BiOCl heterostructure is assigned to the close interface contact and the presence of {001} facets. Mengxia *et al* discovered that the OVs could be changed in the $Bi_7O_9I_3$ flowering microspheres (MSs) by using the procedure that involved the assistance of ionic liquids [15]. These $Bi_7O_9I_3$ MSs, which contain an excessive amount of oxygen, can produce O_2 at a rate of 199.26 μmol g^{-1} h^{-1} when combined with silver nitrate in the role of an electron sacrificial agent. Polyvinylpyrrolidone (PVP) and cetyltrimethylammonium bromide (CTAB) were utilized by Xiong *et al* [16] to adjust the morphology of Bi_3O_4Br so that it took the form of nanoring. Owing to the ring topologies, appropriate band potentials, as well as exposure of the (001) plane, greater mass and carrier adsorption and migration across the nanorings of Bi_3O_4Br were possible. When coupled with $Fe(NO_3)_3$ which acts as an electron-capturing agent, the Bi_3O_4Br nanorings produced O_2 at a rate of 72.54 μmol h^{-1} under the influence of sunlight.

4.4 Photocatalytic hydrogen evolution

Over the past several years, global energy use has expanded substantially, and it is clear that natural resources will not be enough to meet future demands. At the moment, our consumption of energy is dependent on the finite and unsustainable storage of fuels that are derived from fossil sources. At the pace at which energy is being consumed now, fuel reserves will last for up to 100 years. The constant growth in the world's demand for energy has led to a heightened focus on discovering alternative energy sources. As a result, the development of renewable and environmentally friendly forms of energy, such as hydrogen, is essential for industry worldwide. The use of H_2 as an alternative to fossil fuels shows promise. When burned, hydrogen converts into completely safe water vapor and a significant quantity of usable energy [17]. Regarding storing energy, H_2 is chosen over electric batteries because it is lightweight, has a high energy density per unit mass (142.0 MJ kg^{-1}), can be generated on a global scale, and can be kept in storage for a long time [18]. The concept of a 'hydrogen economy' with less reliance on fossil fuels across all industries has been proposed and is gradually becoming a reality [19–25]. However, practically all of the hydrogen utilized on

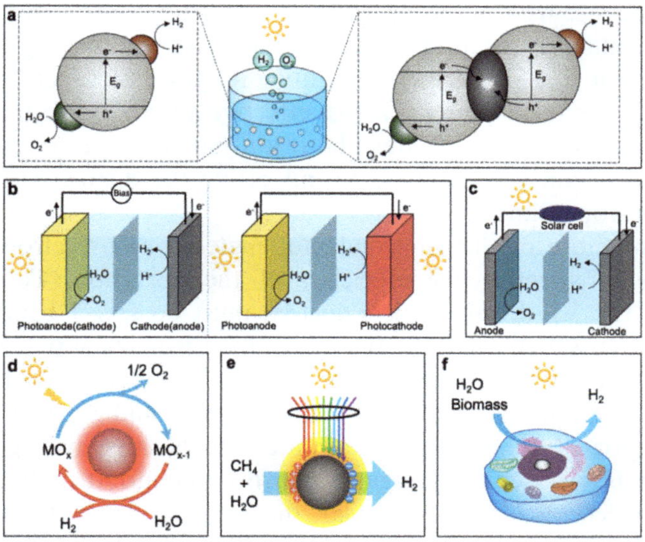

Figure 4.2. (a) PC water splitting, (b) PEC water splitting, (c) PV-EC water splitting, (d) STC water splitting, (e) PTC H_2 production, and (f) PB H_2 production. (Reproduced from [26]. CC BY 4.0.)

Earth comes from the combustion of fossil fuels; 2% of coal and 6% of natural gas are directly converted into hydrogen during this process. These kinds of manufacturing result in annual emissions of 830 megatons of carbon dioxide [18]. As a result, a significant amount of research is being conducted to develop methods for producing 'green' hydrogen without producing carbon-based by-products utilizing renewable energy sources such as wind and solar power. The existing methods for generating H_2 using solar energy are broadly categorized into photoelectrochemical (PEC) water splitting, photobiological (PB) H_2 production, solar thermochemical (STC) water splitting, photovoltaic-electrochemical (PV-EC) water splitting, photothermal catalytic (PTC) production of H_2 from fossil fuels (mainly CH_4), and photocatalytic (PC) water splitting (figure 4.2) [26]. The procedure of water splitting, which uses the abundance of water on our planet, enables the creation of H_2 in a sustainable manner and has several other benefits [27–32]: (i) this method uses pure water, which, in addition to photon energy, is a source of energy; (ii) this process does not produce any pollutants or by-products, making it ecologically benign; and (iii) the creation of hydrogen by photochemical reaction is an effective method for dealing with the seasonal change in solar inflow [33]. The potential of this particular renewable energy technology lies in its capacity to be scaled up to function as a commercial facility dedicated to the production and utilization of H_2. However, to use this technology, there are a number of obstacles that need to be overcome to achieve an effective and economical method of dividing water on a worldwide scale [25]. In contrast to homogeneous counterparts, the utilization of particulate photocatalysts in photocatalytic water splitting, often referred to as heterogeneous catalysis, presents several advantages. One such advantage is the ease with which the catalyst dispersion mixture may be separated following the catalytic reaction. This facilitates the technology's potential for commercialization [2, 25]. Furthermore, utilizing powder suspensions containing particulate photocatalysts presents a cost-effective alternative to the conventional method of fabricating heterogeneous photocatalysts through thin film deposition. This phenomenon can be attributed to mitigating the distinct costs involved in uniformly depositing films.

The band structure of the semiconductor material and electron transport are crucial factors in affecting the efficiency of the operation. To maximize the production of H_2, VL-driven semiconductors must possess a band gap that lies between −1.23 and 3.0 eV. To fulfill the energy requirements of the redox potentials of H_2O, a band gap that is adequate in terms of both conduction and valance bands is required. A semiconductor material with a minimal band gap of 1.23 eV, such as BiOX, is appropriate for the separation of water into its parts, hydrogen and oxygen. The O and Bi faults cause the band gap to narrow and upshifts both the conduction as well as the valence band. In particular, the vacancies in the O atoms lower the energy barrier between O and H, which catalyzes the adsorption of H_2O [34]. While other materials possess a band gap energy similar to that of BiOX, it is essential to understand that their positions in the conduction band as well as the valence band are unique. A higher positive valence band value corresponds to an increased capacity for oxidation, while a more negative conduction band value indicates a stronger aptitude for reduction. Because of its unique layered structure and high level of stability, BiOI has a band gap of 1.78 eV and can thus absorb VL. It is also possible to enhance BiOX photocatalysts by modifying their crystallinity and the shape of their energy bands. Unlike the BiOX, this is not true of other semiconductor materials.

Although the process is feasible, there are still limitations in photocatalysts that prevent them from carrying out the effective breakdown of water. These limitations include photo-corrosion-induced inactivation, poor photostability, and low quantum efficiency of semiconductors with a VL response. The microstructure of the photocatalyst as well as its solubility constant are important factors in determining its photocatalytic stability, together with the pH and the characteristics of the substrate. BiOX, characterized by a significant degree of electron delocalization, is crucial in mitigating the limitations associated with the photocatalytic conversion of water into H_2. This phenomenon relies on the inherent ability of BiOX to undergo photoexcitation upon exposure to sunlight, thereby facilitating the production of conduction band electrons (e^-) and valence band holes (h^+), which subsequently initiate surface reactions.

BiOX has remarkable photocatalytic properties for water purification, water splitting to generate hydrogen, and carbon dioxide conversion due to forming an energy band resulting from p orbitals or s–p hybridized states. Typically, the p electrons within hybridized states possess the ability to lower the conduction band and increase the valence band within the photocatalyst, hence leading to the formation of narrow band gaps. Photocatalysts produced from p-block elements show enhanced visible light PCA due to their small band gaps. This activity might be used to split H_2O into H_2 and convert CO_2 into fossil fuels. Because of this, p-block semiconductors become theoretically possible and attractive for applications in solar light-driven photocatalysis. The formation of a heterojunction can accomplish the generation of H_2 through the use of BiOX [35] or the utilization of defects [36]. Owing to the higher photon absorption as well as improved photoinduced carrier separation efficiency, its PCA can be enhanced. In the process of producing H_2 and O_2 by splitting water and then separating the two gases, it is possible to obtain a higher yield of H_2. Kandi *et al* [37] utilized an *in situ* approach for synthesizing CdS/BiOI composites. They discovered that the BiOI surface demonstrates an increased generation of H_2 gas. The CdS/BiOI compound had the potential to produce 610 μmol of hydrogen within 3 h. They found that increasing the amount of CdS facilitated the enhancement of active sites on the BiOI surface, which improved the light irradiation on that surface. This results in photoexcited electrons, which play a role in releasing hydrogen gas. Sn-doped ZnO/BiOCl heterojunction photocatalysts were fabricated by Guo *et al* [38] using the Z-scheme. Comparing ZBC-S with ZBC, the H_2 evolution rate was calculated as 491.54 and 313.61 μmol g^{-1} h^{-1}, respectively. Adding Eosin Y (EY) at a concentration of 0.434 mg l^{-1} to the composite of ZBC-S resulted in an H_2 generation rate of 4146.77 μmol g^{-1} h^{-1}. This rate was 13 times higher compared to the H_2 yield achieved by ZBC alone. To begin, the ZBC heterojunction was effective in lowering photogenerated carrier recombination. Then, the impurity level formed by doping of Sn decreased the band gap of BiOCl and ZnO to some degree, which increased the ability of the material to absorb light. Additionally, by increasing the concentration of significant charges on the surface of ZBC-S, the sensitization of EY dye increased the light responsiveness of ZBC-S and enhanced the photoefficancy of H_2. Additionally, the same function was discovered in the Schottky junction of BiOBr/C [39]. The absorption of parallel junctions (PJ) exhibited a notable rise in VL due to the advantageous utilization of carbon fiber (CF) photosensitization. On the (001) plane of the BiOBr NSs, the images obtained from Kelvin probe force microscopy and the XPS spectra show a significant number of OVs. Similar to the case of PJs, the (001) crystallographic plane of the BiOBr NSs was in direct contact with the CFs. The electrons that are created by light and captured by OVs are efficiently carried to the CFs due to the presence of a Schottky barrier. Consequently, a significant proportion of the electrons generated through the process of photoexcitation were selectively segregated and subsequently transported to the catalytic sites, CFs, to participate in the reduction of H_2. In addition, the parallel connections' H_2 generation rate reached a maximum of 2850 μmol g^{-1} h^{-1}.

Recent research has shown that using a photocatalyst based on BiOCl for the evolution of H_2 in a system of H_2O, RhB, and methanol is effective [40]. By loading copper phthalocyanine onto the BiOCl, more improvements were made to the effectiveness of the H_2 evolution. The photogenerated hole and electron pair recombination was effectively prevented, and a considerable increase in PCA was accomplished as a result of depositing MnO_x and gold on the (110) and (001) crystal facets of BiOCl, respectively [41]. Despite this, the activity of H_2 evolution was, on average, somewhat modest. Surface oxygen vacancies can be added to BiOCl to increase the activity of H_2 evolution significantly. The glycerol-assisted solvothermal approach was employed to generate BiOCl NSs with surface oxygen vacancies under the irradiation of VL [42]. These NSs demonstrated exhibited enhanced photoactivity for H_2 evolution in comparison to the bulk BiOCl. The heightened level of activity seen can be attributed to the improved efficiency in separating the photoinduced pairs of electron and hole as well as light absorption, which is a direct consequence of the increased spacing between facets and the presence of oxygen vacancies. Homogeneous C doping was proven to significantly raise the ability of BiOCl to produce H_2 when used in conjunction with triethanolamine as a sacrificial electron donor and NiO_x as a co-catalyst [43]. In C-doped {010} faceted BiOCl NSs, the optimum evolution rate of H_2 reached 0.42 μmol g^{-1} h^{-1}. For photocatalytic hydrogen evolution, additional BiOX than BiOCl were also obtained, such as $Bi_{12}O1_{17}Cl_2$ [44] and $Bi_{23}O_{31}Br_{10}$.

4.4.1 Hydrogen generation process

Photophysical (PP) and electrochemical (EC) reactions are the two primary mechanisms involved in the photogeneration of H_2. During the photophysical phenomenon, electrons situated in the valence band undergo photon absorption, resulting in the generation of charge carriers and the migration of electrons towards reaction sites situated in the conduction band. The photochemical process, on the other hand, splits water by redox reactions. Equations (4.1)–(4.5) summarize the entire water splitting reaction and component half-reactions.

Overall water splitting:

$$2H_2O(l) \rightarrow 2H_2(g) + O_2(g) \tag{4.1}$$

H_2 evolution reaction:

$$\text{acidic: } 2H + 2e^- \rightarrow H_2(g) \tag{4.2}$$

$$\text{alkaline: } 2H_2O(l) + 2e^- \rightarrow H_2(g) + 2OH^- \tag{4.3}$$

O_2 evolution reaction:

$$\text{acidic: } 2H_2O(l) + 4h^+ \rightarrow O_2(g)4H \tag{4.4}$$

$$\text{alkaline: } 4OH + 4h^+ \rightarrow O_2(g)2H_2O(l) \tag{4.5}$$

A photocatalyst typically preferentially absorbs photons with energies that are equal to or lower than its band gap energy. In order to cause the ejection of an electron from the photocatalyst's valence band to its conduction band, which results in the formation of holes in the valence band, a photon's energy must be equal to or greater than the photocatalyst's band gap energy. Photon absorption leads to the generation of excited holes and electrons. These electrons, located in the conduction band, can combine with holes by either radiative or nonradiative pathways. This procedure is an inevitable occurrence, independent of the specific semiconductor utilized in the photocatalytic reaction [45–47]. In any photocatalyst, the recombination rate is considered to be the leading factor in evaluating the low quantum efficiency of the catalyst [45, 48–50]. The photoexcited carriers that can avoid recombination will eventually make their way to the surface of the photocatalyst. It is on this surface that the photocatalytic reactions, known as the evolution reaction of hydrogen and oxygen, will take place through the processes of diffusion and electric fields, both of which are associated with the electrolyte or semiconductor interfaces [51]. Excitons are created on timescales generally shorter than one hundred femtoseconds (fs), and their lifetimes are typically measured in the few hundreds of picoseconds (ps). The time it takes for an electron to diffuse through a material is on the order of a few picoseconds, but the time it takes for holes to transition through the material is on the order of 100–300 fs at most [45, 52]. The recombination of photoexcited carriers can also be restricted in semiconductors because to the presence of 'trap sites'. The diffusion coefficients of electrons and holes are decreased when there is a higher number of trap sites present in a catalyst (semiconductor) [53, 54]. The lifespan of electron trapping is only a few microseconds. Still, the processes of hole trapping, recombination, and surface transfer take place substantially more quickly (in the pico- to nano-second range).

The gas evolution rate of H_2 and O_2 is typically expressed using a standardized unit, such as μmol h^{-1} g^{-1}. The PCA is subject to variation based on experimental factors, including the specific catalytic reactor employed, the electrolyte used, and the irradiation light source. Due to the inherent variations, the direct comparison of the PCA among different types of photocatalysts might be challenging. The determination of the quantum yield, which involves making a comparison between the quantity of product (reacted electrons) and the total number of photons that were impacted onto the system, is a workable and generally recognized technique for dealing with this difficulty.

4.5 Photocatalytic nitrogen fixation

4.5.1 Mechanism

The photocatalytic N_2 fixation process is a form of environmentally friendly and non-polluting moderate reaction, devoid of the need for elevated temperatures or high pressures. It has the ability to utilize solar energy to make charge carriers, which it then combines with the protons in water to convert nitrogen dioxide to ammonia (NH_3). Both N_2 oxidation and reduction reactions are possible outcomes of the photoreduction process of nitrogen. In its most basic form, the procedure of N_2 reduction may be broken down into many stages. Initially, the process involves the excitation of electrons in the conduction band through the photoinduced effect caused by solar irradiation. Consequently, this phenomenon leads to the formation of holes in the valence band. Subsequently, a portion of the holes and electrons will undergo recombination with each other. In the meantime, other photoformed holes (h^+) will oxidize the H_2O into O_2 and H^+ (equation (4.6)), and the reduction of N_2 by hot electrons will result in the formation of NH_3 (equation (4.7)). As a consequence, NH_3 may be produced by reacting water and nitrogen dioxide in the presence of sunshine under natural conditions (equation (4.8)) [55]:

$$2H_2O + 4h^+ \rightarrow O_2 + 4H^+ \text{ (1.23 V versus NHE)} \qquad (4.6)$$

$$Z_2 + 6H^+ + 6e^- \rightarrow 2NH_3 \text{ (−0.09 V versus NHE)} \qquad (4.7)$$

$$1/2N_2 + 3/2H_2O \rightarrow NH_3 + 3/4O_2. \qquad (4.8)$$

In the opposite scenario, it was found that the photocatalytic N_2 oxidation process adhered to the photogenerated-hole oxidation mechanism. This was determined by going through the following steps. Initially, when exposed to the irradiation of sunlight, electrons that photogenerated are excited to the conduction band. This leaves holes in the valance band, which is the same result as in the nitrogen reduction reaction situation. After that, photogenerated h^+ oxidize N_2 to NO with H_2O (equation (4.9)), while photoexcited electrons reduce O_2 to H_2O (equation (4.10)), and NO is then further oxidized to the end product nitrates, evolving O_2 and H_2O in the process (equation (4.11)). Nitrate acid may be produced from water, oxygen, and nitrogen under natural conditions with the help of sunlight as an additional source of energy (equation (4.12)) [56]:

$$N_2 + 2H_2O + 4h^+ \rightarrow 2NO + 4H^+ \qquad (4.9)$$

$$O_2 + 4H^+ + 4e^- \rightarrow 2H_2O \qquad (4.10)$$

$$4NO + 3O_2 + 2H_2O \rightarrow 4HNO_3 \qquad (4.11)$$

$$2N_2 + 5O_2 + 2H_2O \rightarrow 4HNO_3. \qquad (4.12)$$

In comparison to the more traditional Haber–Bosch fixation of N_2, fixation of N_2 by photocatalysis is currently regarded as one of the most promising and desirable alternatives. The dissociative mechanism is the theoretical basis for the Haber–Bosch process [57]. During this process, NH_3 is produced when nitrogen and hydrogen ions combine, after the breakdown of the connection between N_2 molecules. To complete this first phase, a large amount of energy must be contributed as the bond energy of N≡N is rather high [58]. Both the associative distal pathway (path 2), as well as the associative alternating pathway (path 1), have been taken into consideration for the associative reduction of N_2; however, the intermediates that are produced by each pathway are unique from one another (figure 4.3). In the alternating route, two N atoms are hydrogenated alternately,

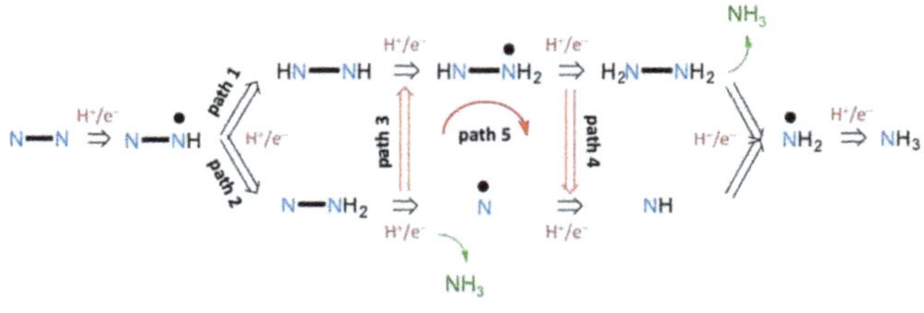

Figure 4.3. Five potential pathways for the photoreduction reaction of N_2. (Reproduced with permission from [56]. Copyright 2019 The Royal Society of Chemistry.)

which results in the formation of hydrazine intermediates after four stages of hydrogenation. However, in the fifth step, only the first NH_3 is liberated from its bound state. In contrast, the distal pathway involves the hydrogenation of a single N atom of N_2 in a total of three stages, which results in the release of the first NH_3, followed by the hydrogenation of the remaining nitride-N in a total of three more steps, which results in the production of the second NH_3 [59]. In the distal associative mechanism, protonation most frequently occurs on the nitrogen atom, which is located a great distance from the catalyst's surface. In contrast, the alternate pathway involves the sequential addition of protons to the two nitrogen atoms of N_2 before the breaking of the N–N bond, which results in one of the nitrogen atoms being transformed into NH_3. The utilization of DFT enables the examination of the intricate mechanisms involved in heterogeneous catalytic N_2 reduction processes in aqueous solutions. It has been observed that the specific routes of these processes are contingent upon the catalyst materials and catalytic systems employed [60, 61].

4.5.2 Role of BiOX for nitrogen fixation

The efficacy of photocatalysts for N_2 reduction has been improved via many studies in the twenty-first century, which have looked into a wide range of potentially effective catalysts [62]. Researchers have discovered that BiOX is suitable as well as potentially useful for the reduction of N_2 through the photocatalysis process. The process that takes place during the photoreduction of N_2 to the BiOX catalyst may be broken down into the following stages (see figure 4.4) [63]: (i) N_2 adsorption takes place at the active site of the BiOX surface for the purpose of N_2 fixation; (ii) BiOX utilizes acquired light energy to generate electrons through the process of photocatalysis; (iii) electrons can recombine with holes, and a portion of both electrons and holes move toward the BiOX surface to engage in the REDOX reaction; and (4)

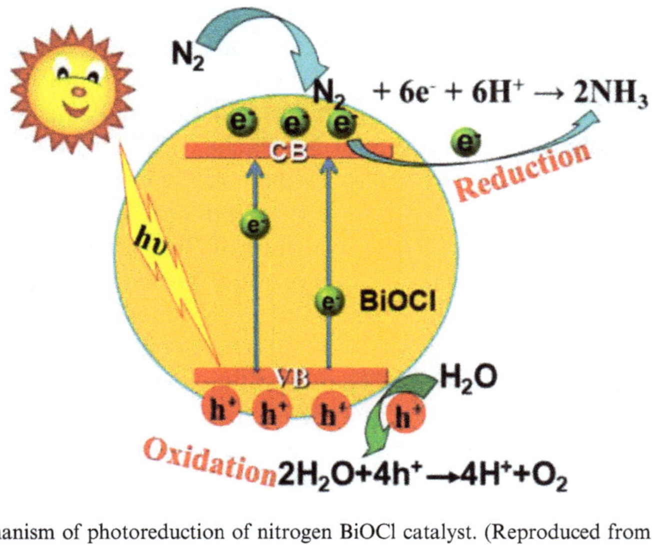

Figure 4.4. Mechanism of photoreduction of nitrogen BiOCl catalyst. (Reproduced from [64]. CC BY 4.0.)

through the holes, water molecules can undergo oxidation, resulting in the production of molecular oxygen, and simultaneously, nitrogen molecules undergo reduction to form NH_3 after a sequence of multi-step injections of photoinduced electrons and protons derived from water.

Pure BiOX showed poor performance in the photocatalytic fixation of N_2. Consequently, numerous investigators have proposed a variety of modification tactics to enhance its photoactivity. For instance, Ai *et al* [65] utilized a non-aqueous sol–gel technique to create BiOBr microspheres to remove NO when the microspheres were irradiated with VL. The synthesized samples demonstrated a higher level of PCA than the Degussa TiO_2 P25 as well as the C-doped TiO_2, in addition to the BiOBr bulk powder. The primary reason for this occurrence can be assigned to an optimal band gap and a distinct hierarchical structure. These characteristics can enhance the absorption capacity of VL and facilitate the rapid diffusion of intermediates, respectively. The synergistic impact of OVs and ultra-thin layer structure was revealed by Xue *et al* [66] to boost the photoreduction N_2 efficiency of BiOBr by about ten times. First, Li *et al* [67] inserted OVs onto the exposed (001) surface of BiOBr NSs. They did this without making use of noble metal promoters or any organic scavengers. It was discovered that OV could adsorb N_2 by working in conjunction with two other OVs that were linked to each other. Synergistically activating N_2 resulted in the development of high-energy intermediates (–HN = NH, $–N_2H$, or $–N_2–$) by the continuous transfer of protons and electrons, which ultimately led to the generation of N_2H_4 or NH_3. The BiOBr sample exhibited an NH_3 evolution rate of 223.3 μmol g^{-1} h^{-1}, surpassing the formation rates of most previously reported single photocatalysts operating under similar conditions. Li *et al* [68] utilized a straightforward and low-temperature thermal treatment technique to produce Bi_5O_7Br nanostructures. The researchers then examined and compared the photocatalytic performance of these nano-structures concerning N_2 fixation. Based on the results obtained, it was seen that the tubular Bi_5O_7Br sample synthesized at a temperature of 40 °C (referred to as Bi_5O_7Br-40) demonstrated the most significant transfer rate of electrons compared to the other samples in the series. Upon exposure to VL photoirradiation, it was observed that a significant quantity of OVs and O_2^- radicals were generated. Furthermore, the photoreduction rate of N_2 reached 12.72 mM g^{-1} h^{-1} following 30 minutes of photoirradiation. The process behind the photocatalytic N_2 fixation performed by Bi_5O_7Br is depicted in figure 4.5. Under the influence of light, the N_2 fixation reaction that takes place on the surface of Bi_5O_7Br may be broken down into two distinct processes: (i) the splitting of water and (ii) the reduction of N_2 to ammonia.

In a recent study, Wang and colleagues [69] conducted research on the synthesis of nanotubes composed of Bi_5O_7Br. Their findings revealed that these nanotubes exhibited an improved rate of photocatalytic N_2 fixing, reaching 1.38 mmol h^{-1} g^{-1}. The VL N_2 fixing rate of H–BiOBr may be as high as 50.8 μmol h^{-1} g^{-1}, according to studies by Bi *et al* [70]. The exceptional photocatalytic efficacy of GQDs/g-C_3N_4/BiOCl with a Z-scheme heterojunction can be attributed to the wide light absorption range of BiOCl/ultrathin g-C_3N_4 binary materials as well as the unique

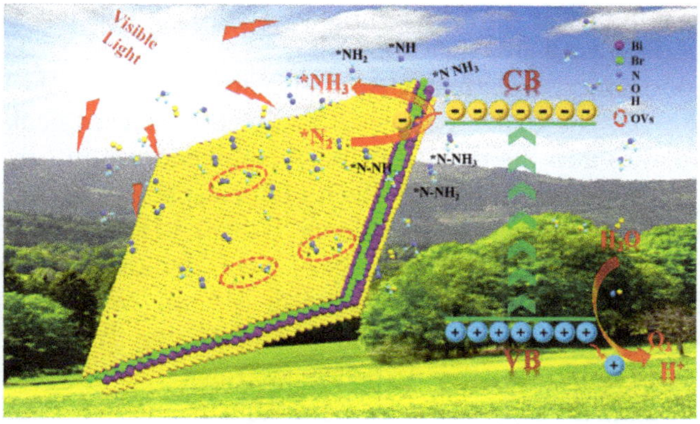

Figure 4.5. Schematic diagram of photoreduction of N_2 catalyzed by Bi_5O_7Br. (Reproduced with permission from [68]. Copyright 2020 American Chemical Society.)

photoelectronic properties of GQDs. After an hour of photocatalytic reaction, the NH_3 yields of GQDs/g-C_3N_4/BiOCl reached 1773.8 μmol g^{-1} h^{-1}, which surpassed those of pure BiOCl and g-C_3N_4 by 5.2 and 7.3 times, respectively [71]. Gao *et al* simultaneously coated the outer and inner surfaces of C_3N_4 nanotubes with BiOBr, which exhibited a flower-like morphology. The successful implementation of this approach resulted in the efficient separation of photoexcited charge carriers, leading to a significant improvement of 13.9 times in the photocatalytic N_2 fixation activity of the BiOBr material [72]. Liu *et al* [73] reported observing a photocatalytic conversion process at the three-phase interface of $Bi_4O_5Br_2$/ZIF-8. This process involved the conversion of water and nitrogen to ammonia and exhibited a rate of 327.338 μmol l^{-1} h^{-1} g^{-1}. In 2016 Li *et al* conducted a pioneering study on the application of BiOCl in the process of photocatalytic fixation of N_2 [74]. This investigation focused on the synthesis of BiOCl NSs and their involvement in the process of proton-assisted electron transfer. The utilization of Bi_2Te_3/BiOCl in the photocatalytic fixation of N_2 was investigated by Rong *et al* [75]. The process was shown to be capable of fixing N_2 at a rate of 315.9 μmol l^{-1} h^{-1}. Guo *et al* [76] successfully fabricated a heterostructure composed of a two-dimensional $ZnIn_2S_4$/BiOCl material. This heterostructure was specifically designed to achieve N_2 fixation, with an impressive rate of 14.6 μmol g^{-1} h^{-1}. According to Wu *et al* [77], the presence of a plethora of OVs and a mostly {001} surface on porous BiOCl microchips doped with Br led to an enhanced capacity for fixation of N_2. BiOI that has been doped or somehow defects has also been utilized for photocatalytic fixation of N_2. The preparation of interstitial BiOI doped with C has been documented by Zeng *et al* [78], and it was shown to have a photocatalytic fixation of N_2 rate of 311 μmol g^{-1} h^{-1}. This rate was approximately 3.7 times higher than that observed for pure BiOI. Lan *et al* [79] conducted a study whereby they synthesized H-Bi_5O_7I microspheres with high concentrations of OVs. The primary aim of this study was to examine the potential of these microspheres for fixation of N_2 under VL conditions.

A rate of 162.48 μmol g^{-1} h^{-1} was measured by H-Bi_5O_7I, which is 7.4- and 2.0-fold higher in comparison with BiOI and Bi_5O_7I.

There is a possibility that the PCA is also influenced by the IEF [80]. The induction of the IEF in BiOX occurs in a direction perpendicular to both the halogen anion planes and the [Bi_2O_2] plane. This phenomenon can be assigned to the layered crystal structure of BiOX, which offers a spacious region that facilitates the polarization of linked orbitals and atoms. Utilizing this particular IEF has the potential to enhance the diffusion of pairs of electrons and holes, resulting in an elevation in carrier mobility and a decrease in the rate of recombination. These effects collectively contribute to enhancing photocatalytic fixation of N_2 efficiency [68, 81–83]. For example, Li *et al* [84] demonstrated that the activity of photocatalysts may be improved by using an approach that involves manipulating the amplitude of the IEF. They prepared single crystal Bi_3O_4Cl NSs with abundant {001} plane exposure. During the photoirradiation process of Bi_3O_4Cl NSs with VL, the movement of electrons from the valence band to the conduction band occurred, leading to the generation of holes in the valence band. The study demonstrated that the polarization-induced IEF along the {001} crystal orientation of the NSs facilitates the migration of charge carriers from the bulk to the surface, aiding further photocatalytic reactions. Consequently, a rise in the level of exposure to the {001} facets resulted in an enhancement of the efficiency in transferring and separating photogenerated pairs of electrons and holes. This, in turn, led to an elevation in the PCA due to the IEF.

4.6 Carbon dioxide photoreduction

The photoreduction of carbon dioxide (CO_2) is a chemical reaction that emulates the biological photosynthesis process, intending to transform CO_2 into environmentally friendly fuels such as methane (CH_4), carbon monoxide (CO), methanol (CH_3OH) as well as formic acid (HCOOH). This chemical process offers a possibility for reducing the dependence on fossil fuels for the generation of energy, while concurrently mitigating the levels of CO_2 present in the Earth's atmosphere [85, 86]. However, since CO_2 has a linear structure, the energy required to rupture the C=O bond is considerably greater compared to the required for breaking the C–H bond, the C–C bond, or the C–O bond [87, 88]. These product generations require various potentials (E_0 V versus NHE at pH 7), as shown in equations (4.13)–(4.19):

$$CO_2 + e^- \rightarrow CO_2^{\bullet -} \quad E_0 = -1.90 \text{ V} \tag{4.13}$$

$$CO_2 + 2H^+ + 2e^- \rightarrow HCOOH + H_2O \quad E_0 = -0.61 \text{ V} \tag{4.14}$$

$$CO_2 + 2H^+ + 2e^- \rightarrow CO + H_2O \quad E_0 = -0.53 \text{ V} \tag{4.15}$$

$$CO_2 + 4H^+ + 4e^- \rightarrow HCHO + H_2O \quad E_0 = -0.48 \text{ V} \tag{4.16}$$

$$CO_2 + 6H^+ + 6e^- \rightarrow CH_3OH + H_2O \quad E_0 = -0.38 \text{ V} \tag{4.17}$$

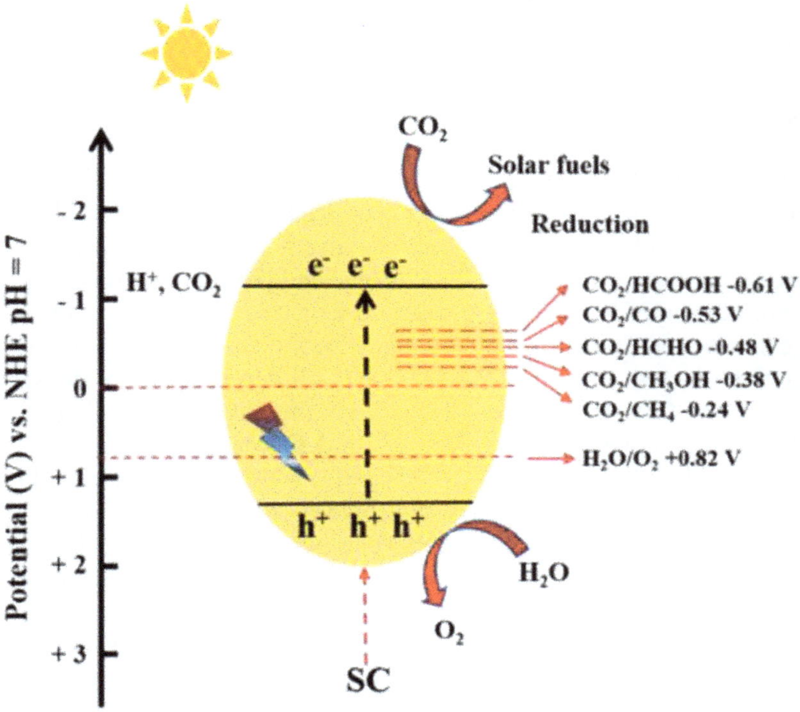

Figure 4.6. Mechanism of PCR. (Reproduced with permission from [91]. Copyright 2022 Elsevier.)

$$CO_2 + 8H^{+}+8e^{-}\rightarrow CH_4 + 2H_2O \; E_0 = -0.\,24 \text{ V} \tag{4.18}$$

$$H_2O + 2h^{+}\rightarrow 1/2O_2 + H^{+} \; E_0 = 0.82 \text{ V}. \tag{4.19}$$

Based on the above information, it can be inferred that the redox potential of −1.90 V (as indicated in equation (4.13)), wherein the CO_2 molecule accepts the initial electron, exhibits the highest degree of potency. The existence of this potential outcome is attributable to the presence of a substantial energy barrier that must be surmounted to induce the curvature of the linear CO_2 molecule on the photocatalyst surface. In many systems for the photoreduction of CO_2, this step is recognized as the limiting stage [89]. The photocatalytic reduction of CO_2 typically involves five steps, as illustrated in figure 4.6 [90]. First, the valence band of the photocatalyst is excited, causing the transfer of an electron (e^-) to its conduction band, resulting in the generation of photoinduced charges. Second, these photoinduced charges have two possible outcomes: they can either migrate and separate to the surface of the photocatalyst to participate in photocatalytic reactions, or they can recombine, leading to the emission of heat or photons. Third, the catalyst's surface facilitates the absorption of CO_2. Forth, the photoinduced electron (e^-) plays a role in converting CO_2 into fuels. Fifth, there is desorption of the product, which refers to the release or removal of a substance from a material or surface.

The efficiency separating of photoexcited charge carriers in photocatalytic CO_2 reduction (PCR) is a crucial metric that significantly affects the selectivity and activity of photocatalysts. However, traditional photocatalysts suffer from the simple recombination of carriers generated by photons, which leads to poor electron transfer efficiency and a performance that is not optimal for CO_2 photoreduction. Recent studies have revealed that the utilization of two-dimensional nanomaterials for the photoreduction of CO_2 results in a significant improvement in photocatalytic carrier separation efficiency. The photocatalysts of the BiOX family feature distinctive layered structures in two dimensions, and they are regarded as potentially useful materials for PCR. The unique layered structures can enhance the separation of charge carriers produced by photons and expedite electron migration toward the surface-active sites. As a result, the PCR reaction might be sped up. Because of this, the reduction of CO_2 by photocatalysts from the BiOX family has become an extremely popular subject [92–94]. An oxygen-deficiency approach for BiOBr NSs that was reported by Kong and co-workers [95] was used to increase the PCA of the NSs. They used an ethylene glycol-assisted solvothermal technique to create oxygen-deficient BiOBr NSs for their research. The synthesized BiOBr NSs showed a significant increase in the production of CH_4 when exposed to VL irradiation. The researchers achieved a cumulative production of 4.86 μmol g^{-1}, whereas the unaltered BiOBr exhibited a lower yield of 1.58 μmol g^{-1} of CH_4. This gain may be ascribed to the fact that the BiOBr NSs could absorb VL better. Specifically, when exposed to synthetic solar light irradiation, oxygen-deficient BiOBr NSs showed a CH_4 yield of 9.58 μmol g^{-1}. This value was 5.7- and 3.2-fold greater than for pristine Degussa P25 and BiOBr, respectively. In light of the fact that the band gap energy of oxygen-deficient BiOBr NSs (2.70 eV) is extremely near to the band gap energy of pure BiOBr (2.85 eV), the increased generation of fuels might primarily be due to numerous variables, including the following. First, the presence of an oxygen-deficient surface significantly enhances the light absorption capabilities of BiOBr, leading to increased production of pairs of electrons and holes. Second, oxygen vacancies facilitate the trapping of photoinduced electrons, thereby enhancing the efficient separation of charge carriers and suppressing their recombination. Lastly, the interaction between oxygen vacancies as well as adsorbed CO_2 molecules potentially gives rise to unforeseen effects, which can further enhance the interfacial charge transfer process [95]. In another study, Wu and colleagues [96] conducted a study to reduce CO_2 emissions. They successfully generated a substantial quantity of OVs on the atomic layers of BiOBr. The resulting CO evolution rate was measured at 87.4 μmol g^{-1} h^{-1}. This rate surpassed that of the bulk and atomic layer BiOBr, exceeding them by factors of 24 and 20, respectively. Furthermore, this achievement represented the highest reported formation rate among most comparable single photocatalysts examined in previous studies. Moreover, it was shown that the atomic layers of oxygen-deficient BiOBr did not exhibit a significant reduction in PCA even after undergoing 60 hours of VL-driven CO_2 reduction. The authors hypothesized a potential reaction pathway (figure 4.7) using *in situ* density functional theory (DFT) and Fourier-transform infrared (FTIR) calculations. Firstly, CO_2 and H_2O molecules are adsorbed onto the surfaces of BiOBr. Following this,

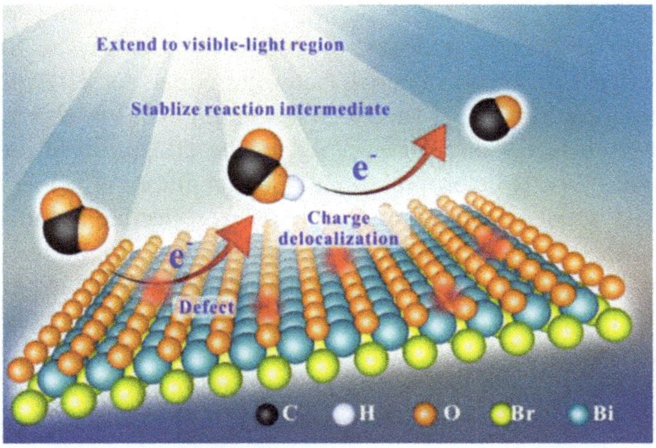

Figure 4.7. The CO_2 photocatalytic reduction into CO over the oxygen-deficient BiOBr atomic layers. (Reproduced with permission from [98]. Copyright 2018 John Wiley and Sons.)

the water molecules (H_2O) that have been adsorbed onto the surface of BiOBr will undergo dissociation, resulting in the formation of hydroxide (OH^-) and hydrogen (H^+) ions. CO_2 will transform CO_2* active species. Furthermore, the CO_2 that is adsorbed onto the surfaces of BiOBr undergoes a reaction with the protons present on the surface, creating an intermediate compound known as COOH*. The production of CO* molecules is achieved through a COOH* intermediate protonation mechanism. The active species of CO* will undergo desorption from the surface of BiOBr, leading to the development of the final CO molecule. The use of OVs in TiO_2@BiOCl resulted in achieving the maximum production rate of CH_4, measuring 168.5 μmol g^{-1} h^{-1}, along with an optimal CH_4 selectivity of 99.4% throughout the process of photocatalytic reduction of CO_2 [97].

To reduce CO_2 emissions, Bai and colleagues [99] created a photocatalyst, Au/BiOI/MnO_x, loaded with dual cocatalysts. In the presence of UV–vis light irradiation, the Au/BiOI/MnO_x sample elucidated a CO production rate of 42.9 μmol h^{-1} g^{-1}. This rate was about seven times greater compared to bare BiOI, which was 6.12 μmol h^{-1} g^{-1}. When subjected to VL, the Au/BiOI/MnO_x sample exhibited an efficient enhancement in the rate of CO, rising from 0.51 μmol h^{-1} g^{-1} for pristine BiOI to 9.76 μmol h^{-1} g^{-1}. After 5 h of photocatalytic reaction, conclusive findings indicated that the total production of CO had attained a value of 169 μmol h^{-1} g^{-1} when the catalyst Au/BiOI/MnO_x was present [99]. This improvement might be attributed to the cocatalysts, which include MnO_x and Au since they not only have the ability to operate as redox-active sites but also can enhance the efficiency of charge carrier separation generated by light. The MnO_x layers and Au nanoparticles are loaded onto the BiOI NSs using photoexcitation as the first step. Next, the photoexcited charger carriers can move to the MnO_x and Au correspondingly, and this transfer mechanism effectively hinders the recombination process of the photoexcited pairs of holes and electrons. Last but not least, the photoexcited electrons can be put to use in the photocatalytic reduction process of CO_2.

To further inhibit the rate at which photogenerated electron–hole pairs recombine and to increase the product efficacy of BiOX photocatalysts, Sun *et al* [100] suggested the formation of an effective heterojunction. As a result, they decided to produce a composite made of In_2O_3/BiOI, leading to a notable enhancement in CO production by a factor of 5.3 compared to the utilization of pure BiOI alone. The formation of a type II heterojunction, which facilitates effective charge transfer and separation at the heterojunction interface, is responsible for this phenomenon. The rate of evolution of photoreduction of CO_2 to CO was found to be highest in the S-scheme heterojunction of AgBr/BiOBr with surface OVs. This particular heterojunction exhibited a rate of 212.6 μmol g^{-1} h^{-1}, which was 9.2 times more than the rate observed in pure BiOBr [101]. In addition, the greatest rate of CO_2 evolution to CH_3OH was reached by the S-scheme heterojunction of BiOBr and pCN, reaching a value of 267.01 μmol g^{-1} h^{-1} [102]. The researchers Jiang *et al* [103] created a one-of-a-kind 2D/2D S-scheme photocatalyst of $CsPbBr_3$/BiOCl by taking advantage of the adequate band structures efficient and interface contact between BiOCl NSs and $CsPbBr_3$. The S-scheme charge transfer mechanism effectively facilitated both separation of charges and maximal redox capacity, resulting in the development of the $CsPbBr_3$/BiOCl heterojunction with a significant driving force for potential CO_2 reduction applications in the future.

4.6.1 Photocatalytic CO_2 reduction: reaction pathways

Glyoxal, carbene, and formaldehyde are the three primary routes postulated for PCR (figure 4.8) [104, 105].

The activation of CO_2 occurs along the pathway leading to the formation of formaldehyde, wherein one of the oxygen atoms binds to the active site of the catalyst. A single electron transfer onto the molecule of CO_2 results in the development of a $CO_2^{\cdot-}$ radical. Subsequently, adding a proton contributes to the development of an intermediate $^{\cdot}$COOH radical. Next, the formation of formic acid takes place when an electron and a proton are successively added to $^{\cdot}$COOH. Next, formaldehyde and water are created when the formic acid takes two protons. According to the existing literature, the primary kinetic obstacle encountered in this pathway is commonly attributed to the photoconversion process of formic acid into formaldehyde [106]. In successive steps of this route, depending on the number of protons and electrons present, it is also possible to make methanol and methane. The synthesis of methane occurs via the $^{\cdot}CH_3$ radical intermediate in this pathway. This pathway can account for the synthesis of methanol, methane, formaldehyde, and formic acid, but it does not generate CO, which is a commonly seen by-product in the process of reducing CO_2.

The utilization of the carbene pathway can lead to the development of CO, which requires the expenditure of two electrons. The CO that is produced can be either a secondary product or an intermediary species that undergoes subsequent reactions to provide methanol or methane. In this instance, the catalyst's active sites establish bonds with the carbon atom of CO_2. The adsorption strength that occurs between the CO as well as the catalyst surface determines whether or not methanol or

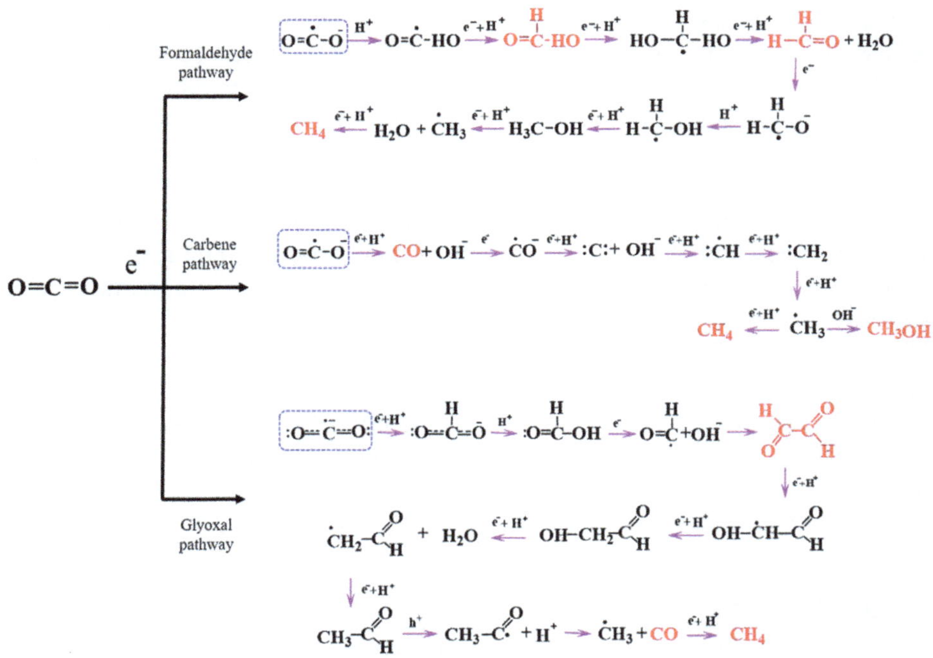

Figure 4.8. General pathways for PCR. (Reproduced from [104]. CC BY 3.0.)

methane is formed via the CO intermediate. The CO molecule has the potential to undergo two distinct processes, fast desorption from the surface or the acquisition of protons and electrons, resulting in the development of the following products. The $^{\cdot}CH_3$ radical that is subsequently created can either undergo a reaction with hydroxide to produce methanol or accept an electron and a proton to produce methane. The carbene pathway has been empirically established as the predominant process for generating CO and other hydrocarbons. The identification of intermediates along this pathway has been facilitated by applying sophisticated analytical methods, including electron paramagnetic resonance [107]. Producing hydrocarbons through the carbene pathway is often known as 'proton-coupled electron transfer'. This reaction is influenced by the catalyst surface's partial electron density and the concentration of available protons, which affects its kinetics. Because the synthesis of CO from CO_2 takes two electrons and two protons, this pathway is difficult. However, manufacturing higher hydrocarbons by 'proton-coupled electron transfer' requires more electrons; for instance, the production of CH_3OH as well as CH_4 requires six and eight electrons, respectively. Although there has been tremendous progress documented in the literature regarding the photocatalytic reduction of CO_2 to afford HCOOH or CO, there is still a gap in the research of the conversion of CO_2 into higher hydrocarbons such as CH_6 and C_2H_5OH with high selectivity and efficiency [108].

The glyoxal pathway is the third possible route for the conversion of CO_2. In the present pathway, the two oxygen atoms of carbon dioxide CO_2 coordinate with the catalytic active site in a bidentate manner, thereby facilitating the formation of a diverse range of products [90]. In the beginning, the $CO_2^{\cdot-}$ radical combines with H^+ to produce a bidentate formate. This bidentate formate then pairs with another H^+ to produce formic acid. Following this, oxygen and electron transfer result in the creation of formyl radicals ($HCO^{\cdot}$). These formyl radicals then dimerize to produce glyoxal before the synthesis of C_2 and C_3 product. In a manner analogous to that of the preceding two processes, the interaction of mCH_3 with a proton results in the synthesis of CH_4 and the removal of CO as a by-product. The formation of C_2 products by the $–CH_3$ radical is highly dependent on other intermediates in the system.

The process of PCR requires numerous phases, intermediates, and distinct products. This makes the process more complicated in obtaining the selectivity of the product, particularly selective access to C_2 compounds that have a greater added value. Even though a substantial amount of headway was achieved in locating the PCR pathway, further theoretical and experimental calculations are still required to clarify the CO_2 reduction pathway as well as its reaction mechanism.

4.7 Organic syntheses

Recent research has shown that photocatalysts based on BiOX can perform selective organic synthesis under moderate circumstances [109]. For instance, the surface-chlorinated BiOBr/TiO_2 materials demonstrated the ability to transform alkanes into oxygenated compounds by activating C–H bonds in the presence of VL irradiation [110]. During the process, BiOBr has a small band gap that was stimulated to form pairs of electrons and holes. These holes were then transported to the valence band of TiO_2 because of the higher positive position of the TiO_2 valance band. The presence of chemisorbed chlorine groups on the catalyst's surface obstructed the available sites, leading to the generation of chlorine radicals. Consequently, these radicals played a crucial role in regulating the activation of C–H bonds. In a recent study, researchers successfully synthesized colloidal ultrathin BiOCl NSs with hydrophobic surface properties. This was achieved through the hydrolysis of $BiCl_3$ in a solution containing octadecylene. The occurrence of this phenomenon was made possible through the *in situ* generation of water resulting from the interaction among oleic acid as well as oleylamine [111]. The BiOCl colloidal ultrathin NSs that were created contained plentiful OVs, which allowed them to demonstrate better PCA for the aerobic oxidation of secondary amines to matching imines under VL irradiation at ambient temperature. Colloidal ultrathin BiOCl NSs demonstrated superior conversion, selectivity, and cycle stability when contrasted with ultrathin NSs of BiOCl generated using a hydrothermal technique. The extraordinary PCA shown in this study can be attributed to the significant absorption of light in the visible range and the pronounced hydrophobic surface characteristics exhibited by the colloidal ultrathin BiOCl NSs. Recent research conducted by the group led by Xie has shown that substantial interactions of electron and hole may be produced in BiOBr because of its limited layered structure. These interactions ultimately lead to powerful excitonic effects [112]. Based on the results obtained from this study, it has been observed that singlet oxygen can be generated in the case of BiOBr with (001) facet exposure, owing to the significant excitonic effect inherent in BiOBr (001), whereas BiOBr (010) produces hydrogen peroxide as the main by-product. For a total of five cycles in a row, the generation of singlet oxygen by BiOBr (001) can promote the mild oxidation of sulfides to the corresponding sulfoxides with high selectivity and powerful photocatalytic activity.

References

[1] Chatterjee K and Skrabalak S E 2021 Durable metal heteroanionic photocatalysts *ACS Appl. Mater. Interfaces* **13** 36670–8

[2] Cui J, Li C and Zhang F 2019 Development of mixed-anion photocatalysts with wide visible-light absorption bands for solar water splitting *ChemSusChem.* **12** 1872–88

[3] Miyoshi A and Maeda K 2021 Recent progress in mixed-anion materials for solar fuel production *Sol. RRL* **5** 2000521

[4] Wang L, Yang G, Wang D, Lu C, Guan W, Li Y, Deng J and Crittenden J 2019 Fabrication of the flower-flake-like $CuBi_2O_4/Bi_2WO_6$ heterostructure as efficient visible-light driven photocatalysts: performance, kinetics and mechanism insight *Appl. Surf. Sci.* **495** 143521

[5] Zeng D, Yang K, Yu C, Chen F, Li X X, Wu Z and Liu H 2018 Phase transformation and microwave hydrothermal guided a novel double Z-scheme ternary vanadate heterojunction with highly efficient photocatalytic performance *Appl. Catal.* B **237** 449–63

[6] Guo Q, Zhou C, Ma Z and Yang X 2019 Fundamentals of TiO_2 photocatalysis: concepts, mechanisms, and challenges *Adv. Mater.* **31** 1901997

[7] Xiang Y, Ju P, Wang Y, Sun Y, Zhang D and Yu J 2016 Chemical etching preparation of the Bi_2WO_6/BiOI p–n heterojunction with enhanced photocatalytic antifouling activity under visible light irradiation *Chem. Eng. J.* **288** 264–75

[8] Angaiah S, Arunachalam S, Murugadoss V and Vijayakumar G 2019 A facile polyvinyl-pyrrolidone assisted solvothermal synthesis of zinc oxide nanowires and nanoparticles and their influence on the photovoltaic performance of dye sensitized solar cell *ES Energy Environ.* **4** 59–65

[9] Zhou L, Song W, Chen Z and Yin G 2013 Degradation of organic pollutants in wastewater by bicarbonate-activated hydrogen peroxide with a supported cobalt catalyst *Environ. Sci. Technol.* **47** 3833–9

[10] Bai S, Li X, Kong Q, Long R, Wang C, Jiang J and Xiong Y 2015 Toward enhanced photocatalytic oxygen evolution: synergetic utilization of plasmonic effect and Schottky junction via interfacing facet selection *Adv. Mater.* **27** 3444–52

[11] Jiang Z, Liu Y, Jing T, Huang B, Wang Z, Zhang X, Qin X and Dai Y 2015 One-pot solvothermal synthesis of S doped BiOCl for solar water oxidation *RSC Adv.* **5** 47261–4

[12] Li J, Cai L, Shang J, Yu Y and Zhang L 2016 Giant enhancement of internal electric field boosting bulk charge separation for photocatalysis *Adv. Mater.* **28** 4059–64

[13] Wu W, Zhang Z, Di J and Zhao W 2019 Improved solar energy photoactivity over defective BiOBr ultrathin nanosheets towards pollutant removal and oxygen evolution *ChemNanoMat* **5** 215–23

[14] Ning S, Shi X, Zhang H, Lin H, Zhang Z, Long J, Li Y and Wang X 2019 Reconstructing dual-induced {0 0 1} facets bismuth oxychloride nanosheets heterostructures: an effective strategy to promote photocatalytic oxygen evolution *Sol. RRL* **3** 1900059

[15] Ji M, Chen R, Di J, Liu Y, Li K, Chen Z, Xia J and Li H 2019 Oxygen vacancies modulated Bi-rich bismuth oxyiodide microspheres with tunable valence band position to boost the photocatalytic activity *J. Colloid Interface Sci.* **533** 612–20

[16] Xiong X, Zhou T, Liu X, Ding S and Hu J 2017 Surfactant-mediated synthesis of single-crystalline Bi_3O_4Br nanorings with enhanced photocatalytic activity *J. Mater. Chem.*A **5** 15706–13

[17] Dai Y *et al* 2018 Efficient solar-driven hydrogen transfer by bismuth-based photocatalyst with engineered basic sites *J. Am. Chem. Soc.* **140** 16711–9

[18] Gordon M N, Chatterjee K, Beena N C and Skrabalak S E 2022 Sustainable production of layered bismuth oxyhalides for photocatalytic H_2 production *ACS Sustain. Chem. Eng.* **10** 15622–41

[19] Guy K W A 2000 The hydrogen economy *Process Saf. Environ. Prot.* **78** 324–7

[20] Staffell I, Scamman D, Velazquez Abad A, Balcombe P, Dodds P E, Ekins P, Shah N and Ward K R 2019 The role of hydrogen and fuel cells in the global energy system *Energy Environ. Sci.* **12** 463–91

[21] van Renssen S 2020 The hydrogen solution? *Nat. Clim. Chang.* **10** 799–801

[22] Rosen M A and Koohi-Fayegh S 2016 The prospects for hydrogen as an energy carrier: an overview of hydrogen energy and hydrogen energy systems *Energy, Ecol. Environ.* **1** 10–29

[23] Brandon N P and Kurban Z 2017 Clean energy and the hydrogen economy *Philos. Trans. R. Soc.* A **375**

[24] Eljack F and Kazi M K 2020 Prospects and challenges of green hydrogen economy via multi-sector global symbiosis in Qatar *Front. Sustain.* **1**

[25] Wang Q and Domen K 2020 Particulate photocatalysts for light-driven water splitting: mechanisms, challenges, and design strategies *Chem. Rev.* **120** 919–85

[26] Song H, Luo S, Huang H, Deng B and Ye J 2022 Solar-driven hydrogen production: recent advances, challenges, and future perspectives *ACS Energy Lett.* **7** 1043–65

[27] Chu S and Majumdar A 2012 Opportunities and challenges for a sustainable energy future *Nature* **488** 294–303

[28] Huang C, Wen Y, Ma J, Dong D, Shen Y, Liu S, Ma H and Zhang Y 2021 Unraveling fundamental active units in carbon nitride for photocatalytic oxidation reactions *Nat. Commun.* **12** 320

[29] Zhou G *et al* 2018 Half-metallic carbon nitride nanosheets with micro grid mode resonance structure for efficient photocatalytic hydrogen evolution *Nat. Commun.* **9** 3366

[30] Ikram M, Umar E, Raza A, Haider A, Naz S, Ul-Hamid A, Haider J, Shahzadi I, Hassan J and Ali S 2020 Dye degradation performance, bactericidal behavior and molecular docking analysis of Cu-doped TiO_2 nanoparticles *RSC Adv.* **10** 24215–33

[31] Kosco J *et al* 2020 Enhanced photocatalytic hydrogen evolution from organic semi-conductor heterojunction nanoparticles *Nat. Mater.* **19** 559–65

[32] Raza A, Qumar U, Hassan J, Ikram M, Ul-Hamid A, Haider J, Imran M and Ali S 2020 A comparative study of dirac 2D materials, TMDCs and 2D insulators with regard to their structures and photocatalytic/sonophotocatalytic behavior *Appl. Nanosci.* **10** 3875–99

[33] Deng F, Zou J P, Zhao L N, Zhou G, Luo X B and Luo S L 2018 Nanomaterial-based photocatalytic hydrogen production *Nanomaterial-Based Photocatalytic Hydrogen Production* (Amsterdam: Elsevier) pp 59–82

[34] Liu Q C, Ma D K, Hu Y Y, Zeng Y W and Huang S M 2013 Various bismuth oxyiodide hierarchical architectures: alcohothermal-controlled synthesis, photocatalytic activities, and adsorption capabilities for phosphate in water *ACS Appl. Mater. Interfaces* **5** 11927–34

[35] Mei F, Zhang J, Dai K, Zhu G and Liang C 2019 A Z-scheme Bi_2MoO_6/CdSe-diethylenetriamine heterojunction for enhancing photocatalytic hydrogen production activity under visible light *Dalton Trans.* **48** 1067–74

[36] Park H, Bak A, Ahn Y Y, Choi J and Hoffmannn M R 2012 Photoelectrochemical performance of multi-layered BiO_x–TiO_2/Ti electrodes for degradation of phenol and production of molecular hydrogen in water *J. Hazard. Mater.* **211–12** 47–54

[37] Kandi D, Martha S, Thirumurugan A and Parida K M 2017 Modification of BiOI microplates with CdS QDs for enhancing stability, optical property, electronic behavior toward Rhodamine B decolorization, and photocatalytic hydrogen evolution *J. Phys. Chem.* C **121** 4834–49

[38] Guo Y, Qi C, Lu B and Li P 2022 Enhanced hydrogen production from water splitting by Sn-doped ZnO/BiOCl photocatalysts and Eosin Y sensitization *Int. J. Hydrogen Energy* **47** 228–41

[39] Zheng X, Feng L, Dou Y, Guo H, Liang Y, Li G, He J, Liu P and He J 2021 High carrier separation efficiency in morphology-controlled BiOBr/C Schottky junctions for photocatalytic overall water splitting *ACS Nano* **15** 13209–19

[40] Zhang L, Wang W, Sun S, Sun Y, Gao E and Xu J 2013 Water splitting from dye wastewater: a case study of BiOCl/copper(II) phthalocyanine composite photocatalyst *Appl. Catal.* B **132–3** 315–20

[41] Zhang L, Wang W, Sun S, Jiang D and Gao E 2015 Selective transport of electron and hole among {001} and {110} facets of BiOCl for pure water splitting *Appl. Catal.* B **162** 470–4

[42] Ye L, Jin X, Leng Y, Su Y, Xie H and Liu C 2015 Synthesis of black ultrathin BiOCl nanosheets for efficient photocatalytic H_2 production under visible light irradiation *J. Power Sources* **293** 409–15

[43] Li J, Zhao K, Yu Y and Zhang L 2015 Facet-level mechanistic insights into general homogeneous carbon doping for enhanced solar-to-hydrogen conversion *Adv. Funct. Mater.* **25** 2189–201

[44] Li J, Zhan G, Yu Y and Zhang L 2016 Superior visible light hydrogen evolution of Janus bilayer junctions via atomic-level charge flow steering *Nat. Commun.* **7** 11480

[45] Rahman M Z, Kibria M G and Mullins C B 2020 Metal-free photocatalysts for hydrogen evolution *Chem. Soc. Rev.* **49** 1887–931

[46] Chen X, Shen S, Guo L and Mao S S 2010 Semiconductor-based photocatalytic hydrogen generation *Chem. Rev.* **110** 6503–70

[47] Ravelli D, Dondi D, Fagnoni M and Albini A 2009 Photocatalysis. A multi-faceted concept for green chemistry *Chem. Soc. Rev.* **38** 1999–2011

[48] Ohtani B 2010 Photocatalysis A to Z—what we know and what we do not know in a scientific sense *J. Photochem. Photobiol.* C **11** 157–78

[49] Ikram M, Hassan J, Imran M, Haider J, Ul-Hamid A, Shahzadi I, Ikram M, Raza A, Qumar U and Ali S 2020 2D chemically exfoliated hexagonal boron nitride (hBN) nanosheets doped with Ni: synthesis, properties and catalytic application for the treatment of industrial wastewater *Appl. Nanosci.* **10** 3525–8

[50] Ikram M, Hassan J, Raza A, Haider A, Naz S, Ul-Hamid A, Haider J, Shahzadi I, Qamar U and Ali S 2020 Photocatalytic and bactericidal properties and molecular docking analysis

of TiO_2 nanoparticles conjugated with Zr for environmental remediation *RSC Adv.* **10** 30007–24
[51] Hisatomi T, Takanabe K and Domen K 2015 Photocatalytic water-splitting reaction from catalytic and kinetic perspectives *Catal. Lett.* **145** 95–108
[52] Kroeze J E, Savenije T J and Warman J M 2004 Electrodeless determination of the trap density, decay kinetics, and charge separation efficiency of dye-sensitized nanocrystalline TiO_2 *J. Am. Chem. Soc.* **126** 7608–18
[53] Bisquert J 2008 Interpretation of electron diffusion coefficient in organic and inorganic semiconductors with broad distributions of states *Phys. Chem. Chem. Phys.* **10** 3175–94
[54] Shen M and Henderson M A 2011 Identification of the active species in photochemical hole scavenging reactions of methanol on TiO_2 *J. Phys. Chem. Lett.* **2** 2707–10
[55] Vu M H, Sakar M and Do T O 2018 Insights into the recent progress and advanced materials for photocatalytic nitrogen fixation for ammonia (NH_3) production *Catalysts* **8** 621
[56] Cheng M, Xiao C and Xie Y 2019 Photocatalytic nitrogen fixation: the role of defects in photocatalysts *J. Mater. Chem.* A **7** 19616–33
[57] Tan H, Zhao Z, Bin Zhu W, Coker E N, Li B, Zheng M, Yu W, Fan H and Sun Z 2014 Oxygen vacancy enhanced photocatalytic activity of pervoskite $SrTiO_3$ *ACS Appl. Mater Interfaces* **6** 19184–90
[58] John J, Lee D K and Sim U 2019 Photocatalytic and electrocatalytic approaches towards atmospheric nitrogen reduction to ammonia under ambient conditions *Nano Converg.* **6** 15
[59] Hou T *et al* 2019 Operando oxygen vacancies for enhanced activity and stability toward nitrogen photofixation *Adv. Energy Mater.* **9** 1902319
[60] Guo C, Ran J, Vasileff A and Qiao S Z 2018 Rational design of electrocatalysts and photo(electro)catalysts for nitrogen reduction to ammonia (NH_3) under ambient conditions *Energy Environ. Sci.* **11** 45–56
[61] Zhang L, Wang W, Jiang D, Gao E and Sun S 2015 Photoreduction of CO_2 on BiOCl nanoplates with the assistance of photoinduced oxygen vacancies *Nano Res.* **8** 821–31
[62] Medford A J and Hatzell M C 2017 Photon-driven nitrogen fixation: current progress, thermodynamic considerations, and future outlook *ACS Catal.* **7** 2624–43
[63] Chen X, Li N, Kong Z, Ong W J and Zhao X 2018 Photocatalytic fixation of nitrogen to ammonia: state-of-the-art advancements and future prospects *Mater. Horizons* **5** 9–27
[64] Yuan Z and Jiang Z 2023 Applications of BiOX in the photocatalytic reactions *Molecules* **28** 4400
[65] Ai Z, Ho W, Lee S and Zhang L 2009 Efficient photocatalytic removal of NO in indoor air with hierarchical bismuth oxybromide nanoplate microspheres under visible light *Environ. Sci. Technol.* **43** 4143–50
[66] Xue X *et al* 2018 Oxygen vacancy engineering promoted photocatalytic ammonia synthesis on ultrathin two-dimensional bismuth oxybromide nanosheets *Nano Lett.* **18** 7372–7
[67] Li H, Shang J, Ai Z and Zhang L 2015 Efficient visible light nitrogen fixation with BiOBr nanosheets of oxygen vacancies on the exposed {001} facets *J. Am. Chem. Soc.* **137** 6393–9
[68] Li P *et al* 2020 Visible-light-driven nitrogen fixation catalyzed by Bi_5O_7Br nanostructures: enhanced performance by oxygen vacancies *J. Am. Chem. Soc.* **142** 12430–9
[69] Wang S, Hai X, Ding X, Chang K, Xiang Y, Meng X, Yang Z, Chen H and Ye J 2017 Light-switchable oxygen vacancies in ultrafine Bi_5O_7Br nanotubes for boosting solar-driven nitrogen fixation in pure water *Adv. Mater.* **29** 1701774

[70] Bi Y, Wang Y, Dong X, Zheng N, Ma H and Zhang X 2018 Efficient solar-driven conversion of nitrogen to ammonia in pure water: via hydrogenated bismuth oxybromide *RSC Adv.* **8** 21871–8

[71] Cai J, Maimaitizi H, Okitsu K, Tursun Y and Abulizi A 2022 Z-type heterojunction of graphene quantum dots/g-C_3N_4/BiOCl with excellent photocatalytic performance for nitrogen fixation *Int. J. Energy Res.* **46** 12147–59

[72] Gao K, Zhang C, Zhu H, Xia J, Chen J, Xie F, Zhao X, Tang Z and Wang X 2023 Unique tubular BiOBr/g-C_3N_4 heterojunction with efficient separation of charge carriers for photocatalytic nitrogen fixation *Chemistry* **29** e202300616

[73] Liu J, Li R, Zu X, Zhang X, Wang Y, Wang Y and Fan C 2019 Photocatalytic conversion of nitrogen to ammonia with water on triphase interfaces of hydrophilic–hydrophobic composite $Bi_4O_5Br_2$/ZIF-8 *Chem. Eng. J.* **371** 796–803

[74] Li H, Shang J, Shi J, Zhao K and Zhang L 2016 Facet-dependent solar ammonia synthesis of BiOCl nanosheets via a proton-assisted electron transfer pathway *Nanoscale* **8** 1986–93

[75] Rong X, Mao Y, Xu J, Zhang X, Zhang L, Zhou X, Qiu F and Wu Z 2018 Bi_2Te_3 sheet contributing to the formation of flower-like BiOCl composite and its N_2 photofixation ability enhancement *Catal. Commun.* **116** 16–9

[76] Guo L, Han X, Zhang K, Zhang Y, Zhao Q, Wang D and Fu F 2019 *In-situ* construction of 2D/2D $ZnIn_2S_4$/BiOCl heterostructure with enhanced photocatalytic activity for N_2 fixation and phenol degradation *Catalysts* **9** 729

[77] Wu D *et al* 2019 Br doped porous bismuth oxychloride micro-sheets with rich oxygen vacancies and dominating {0 0 1} facets for enhanced nitrogen photo-fixation performances *J. Colloid Interface Sci.* **556** 111–9

[78] Zeng L, Zhe F, Wang Y, Zhang Q, Zhao X, Hu X, Wu Y and He Y 2019 Preparation of interstitial carbon doped BiOI for enhanced performance in photocatalytic nitrogen fixation and methyl orange degradation *J. Colloid Interface Sci.* **539** 563–74

[79] Lan M, Zheng N, Dong X, Hua C, Ma H and Zhang X 2020 Bismuth-rich bismuth oxyiodide microspheres with abundant oxygen vacancies as an efficient photocatalyst for nitrogen fixation *Dalton Trans.* **49** 9123–9

[80] Chen P, Liu H, Cui W, Lee S C, Wang L and Dong F 2020 Bi-based photocatalysts for light-driven environmental and energy applications: structural tuning, reaction mechanisms, and challenges *EcoMat* **2** e12047

[81] Ye L, Su Y, Jin X, Xie H and Zhang C 2014 Recent advances in BiOX (X = Cl, Br and I) photocatalysts: synthesis, modification, facet effects and mechanisms *Environ. Sci. Nano.* **1** 90–112

[82] He R, Cao S, Zhou P and Yu J 2014 Recent advances in visible light Bi-based photocatalysts *Cuihua Xuebao/Chin. J. Catal* **35** 989–1007

[83] Xu K, Wang L, Xu X, Dou S X, Hao W and Du Y 2019 Two dimensional bismuth-based layered materials for energy-related applications *Energy Storage Mater.* **19** 446–63

[84] Li J, Zhang L, Li Y and Yu Y 2014 Synthesis and internal electric field dependent photoreactivity of Bi_3O_4Cl single-crystalline nanosheets with high {001} facet exposure percentages *Nanoscale* **6** 167–71

[85] Moustakas N G and Strunk J 2018 Photocatalytic CO_2 reduction on TiO_2-based materials under controlled reaction conditions: systematic insights from a literature study *Chemistry* **24** 12739–46

[86] Corma A and Garcia H 2013 Photocatalytic reduction of CO_2 for fuel production: possibilities and challenges *J. Catal.* **308** 168–75
[87] Sun Z, Wang H, Wu Z and Wang L 2018 g-C_3N_4 based composite photocatalysts for photocatalytic CO_2 reduction *Catal. Today* **300** 160–72
[88] Bi Z X, Guo R T, Hu X, Wang J, Chen X and Pan W G 2022 Research progress on photocatalytic reduction of CO_2 based on LDH materials *Nanoscale* **14** 3367–86
[89] Zhang Y, Xia B, Ran J, Davey K and Qiao S Z 2020 Atomic-level reactive sites for semiconductor-based photocatalytic CO_2 reduction *Adv. Energy Mater.* **10** 1903879
[90] Ong W J, Putri L K and Mohamed A R 2020 Rational design of carbon-based 2D nanostructures for enhanced photocatalytic CO_2 reduction: a dimensionality *Persp. Chem.* A **26** 9710–48
[91] Hu X, Guo R T, Chen X, Bi Z X, Wang J and Pan W G 2022 Bismuth-based Z-scheme structure for photocatalytic CO_2 reduction: a review *J. Environ. Chem. Eng.* **10** 108582
[92] Zhao X, Xia Y, Li H, Wang X, Wei J, Jiao X and Chen D 2021 Oxygen vacancy dependent photocatalytic CO_2 reduction activity in liquid-exfoliated atomically thin BiOCl nanosheets *Appl. Catal.* B **297** 120426
[93] Zhu J Y, Li Y P, Wang X J, Zhao J, Wu Y S and Li F T 2019 Simultaneous phosphorylation and Bi modification of BiOBr for promoting photocatalytic CO_2 reduction *ACS Sustain. Chem. Eng.* **7** 14953–61
[94] Sun N, Zhou M, Ma X, Cheng Z, Wu J, Qi Y, Sun Y, Zhou F, Shen Y and Lu S 2022 Self-assembled spherical In_2O_3/BiOI heterojunctions for enhanced photocatalytic CO_2 reduction activity *J. CO_2 Util.* **65** 102220
[95] Kong X Y, Lee W P C, Ong W J, Chai S P and Mohamed A R 2016 Oxygen-deficient BiOBr as a highly stable photocatalyst for efficient CO_2 reduction into renewable carbon-neutral fuels *ChemCatChem.* **8** 3074–81
[96] Wu J *et al* 2018 Efficient visible-light-driven CO_2 reduction mediated by defect-engineered BiOBr atomic layers *Angew. Chem.* **130** 8855–9
[97] Li R, Luan Q, Dong C, Dong W, Tang W, Wang G and Lu Y 2021 Light-facilitated structure reconstruction on self-optimized photocatalyst TiO_2@BiOCl for selectively efficient conversion of CO_2 to CH_4 *Appl. Catal.* B***286***
[98] Wu J *et al* 2018 Efficient visible-light-driven CO_2 reduction mediated by defect-engineered BiOBr atomic layers *Angew. Chem., Int. Ed.* **57** 8719–23
[99] Bai Y, Ye L, Wang L, Shi X, Wang P and Bai W 2016 A dual-cocatalyst-loaded Au/BiOI/MnO:X system for enhanced photocatalytic greenhouse gas conversion into solar fuels *Environ. Sci. Nano.* **3** 902–9
[100] Wang R, Shen J, Sun K, Tang H and Liu Q 2019 Enhancement in photocatalytic activity of CO_2 reduction to CH_4 by 0D/2D Au/TiO_2 plasmon heterojunction *Appl. Surf. Sci.* **493** 1142–9
[101] Miao Z, Wang Q, Zhang Y, Meng L and Wang X 2022 *In situ* construction of S-scheme AgBr/BiOBr heterojunction with surface oxygen vacancy for boosting photocatalytic CO_2 reduction with H_2O *Appl. Catal.* B **301** 120802
[102] zhang T, Maihemllti M, Okitsu K, Talifur D, Tursun Y and Abulizi A 2021 *In situ* self-assembled S-scheme BiOBr/pCN hybrid with enhanced photocatalytic activity for organic pollutant degradation and CO_2 reduction *Appl. Surf. Sci.* **556** 149828

[103] Jiang Y, Wang Y, Zhang Z, Dong Z and Xu J 2022 2D/2D $CsPbBr_3$/BiOCl heterojunction with an S-scheme charge transfer for boosting the photocatalytic conversion of CO_2 *Inorg. Chem.* **61** 10557–66

[104] Gong E, Ali S, Hiragond C B, Kim H S, Powar N S, Kim D and Kim H 2021 Solar fuels: research and development strategies to accelerate photocatalytic CO_2 conversion into hydrocarbon fuels *Energy Environ. Sci.* **15** 880–937

[105] Kong T, Jiang Y and Xiong Y 2020 Photocatalytic CO_2 conversion: what can we learn from conventional CO:X hydrogenation? *Chem. Soc. Rev.* **49** 6579–91

[106] Ji Y and Luo Y 2016 Theoretical study on the mechanism of photoreduction of CO_2 to CH_4 on the anatase TiO_2(101) surface *ACS Catal.* **6** 2018–25

[107] Fu J, Jiang K, Qiu X, Yu J and Liu M 2020 Product selectivity of photocatalytic CO_2 reduction reactions *Mater. Today* **32** 222–43

[108] Chang X, Wang T and Gong J 2016 CO_2 photo-reduction: insights into CO_2 activation and reaction on surfaces of photocatalysts *Energy Environ. Sci.* **9** 2177–96

[109] Ding L, Chen H, Wang Q, Zhou T, Jiang Q, Yuan Y, Li J and Hu J 2016 Synthesis and photocatalytic activity of porous bismuth oxychloride hexagonal prisms *Chem. Commun.* **52** 994–7

[110] Yuan R *et al* 2013 Chlorine-radical-mediated photocatalytic activation of C–H bonds with visible light *Angew. Chem., Int. Ed.* **52** 1035–9

[111] Wu Y, Yuan B, Li M, Zhang W H, Liu Y and Li C 2015 Well-defined BiOCl colloidal ultrathin nanosheets: synthesis, characterization, and application in photocatalytic aerobic oxidation of secondary amines *Chem. Sci.* **6** 1873–8

[112] Wang H, Chen S, Yong D, Zhang X, Li S, Shao W, Sun X, Pan B and Xie Y 2017 Giant electron–hole interactions in confined layered structures for molecular oxygen activation *J. Am. Chem. Soc.* **139** 4737–42

Chapter 5

Water purification applications

The phenomenon of water contamination is increasing steadily, presenting a significant treat to the entire spectrum of living organisms. The primary contributors to water contamination include many substances such as dyes, antibiotics, chemical waste, and bacteria. Various approaches have been explored by researchers to address this issue, among which the photocatalysis process utilizing BiOX-based materials has garnered substantial attention. This chapter presents a thorough examination of the research conducted on BiOX photocatalysts concerning their efficacy in the decolorization of organic dyes, antibiotics, as well as water disinfection. Furthermore, the mechanisms of pollutant breakdown and disinfection by BiOX-based materials are discussed. In addition, a succinct examination of water contaminants and their impact on human existence is presented. Finally, we present a study of the reduction of heavy metals and the underlying mechanism facilitated by photocatalysts based on BiOX.

5.1 Water pollution

The rapid pace of industrialization has contributed to a wide variety of pollution problems in the natural environment. Wastewater is being released into ordinary water sources more frequently, and it may contain radioactive nuclides, organic pollutants, and heavy metals. Both organic pollutants and hazardous materials pose a considerable risk of pollution of fresh water, and they have a propensity to bioaccumulate and cause repercussions for human health. As a result, the elimination of harmful contaminants as well as their identification are crucial. Organic pollutants are mostly made up of hydrogen and carbon, along with minute amounts of various additional elements. The majority of hazardous chemicals, including dye compounds, per fluorinated compounds, polychlorinated biphenyls, hexachlorobenzene, organochlorine insecticides, dibenzofurans, dichlorodiphenyltrichloroethane (DDT), furans, dioxins, polynuclear aromatic hydrocarbons, and many more, are included in organic pollutants. Pigments for dyes that include azo linkages (–N– –N–)

doi:10.1088/978-0-7503-5934-4ch5

are commonly utilized in industrial processes, and as a result, these processes generate potentially hazardous by-products. The generation of these by-products can occur through processes such as hydrolysis, oxidation, and chemical reactions that occur within wastewater or aquatic environments. Chemical industries, including those involved in the production of paper, food, plastic, leather, and ceramic materials, serve as significant contributors of dyes and various organic pollutants. Organic pollutants enter the body mostly through consuming various forms of meat, fish, and dairy products. The ingestion of pollutants puts humans at high risk for a variety of diseases and conditions, including cancer, the disruption of thyroid function and sex hormones, the suppression of the immune system, chronic diseases, hypertension, neurological problems, diabetes, and cardiovascular diseases.

In addition to biological contaminants, wastewater often contains inorganic pollutants such as heavy metals. In the past few decades, research on the removal of organic pollutants as well as the elimination of heavy metals including chromium (Cr), cadmium (Cd), arsenic (Ar), nickel (Ni), lead (Pb), and mercury (Hg) has garnered a substantial amount of interest. Heavy metals are incapable of breaking down and can remain viable for generations. The waste from manufacturers, excessive use of fertilizers and pesticides, and common household items are the biggest contributors to the presence of heavy metals. The build-up of heavy metals has a contaminating effect on water and land, which in turn has an impact on the environment and human existence. Bulk elements are necessary for humans and include the vital metals magnesium, calcium, potassium, and sodium. Bulk elements are also known as trace elements. Trace elements such as Zn, Mo, Co, and Cu are necessary for protein enzymes in diets such as nitrogen-fixing organisms and vitamin B12. Toxic effects are caused by these trace metals and other heavy metals when they are in larger amounts. Children who are exposed to heavy metals have symptoms such as bloody diarrhea, renal failure, and a decrease in their intelligence quotient and intellectual function. Arsenic (Ar), cadmium, and chromium have all been linked to disruptions in DNA synthesis as well as DNA damage and neuropsychiatric problems [1].

There is the possibility that pharmaceutical residues, such as antibiotics, are present in the wastewater that is discharged from medical facilities and manufacturing facilities that produce pharmaceuticals. Antibiotics have been the subject of much research regarding the harm they pose to aquatic creatures. The biological contamination of water might also result in the transmission of waterborne illnesses such as jaundice, dysentery, typhoid, cholera, and diarrhea, amongst others. The World Health Organization estimates that 4 850 000 people die every year as a direct result of diarrhea. Because of this, having access to clean water is a necessity for human beings and is necessary to meet the requirements of human activities [2]. It should come as no surprise that disinfecting polluted water is the most effective strategy for addressing the shortage of clean water. Around the world, people use several techniques to clean water, including electrochemical approaches, dialysis, Fenton processes, reverse osmosis, ultrafiltration, sono-catalysis, adsorption, phytoremediation, bioremediation, and precipitation–coagulation [3]. Traditional water treatment methods, on the other hand, have several drawbacks, such as the fact that they consume a lot of energy, require a lot of time to complete, involve costly equipment and chemicals, and have the potential to cause sludge build-up and secondary pollution. This makes it challenging to develop practical applications that are both highly economical and highly efficient. For all of these reasons, it is necessary to research and create water purification methods that are both economical and efficient in their use of energy. In this setting, heterogeneous photocatalysis is rapidly becoming a viable option for treating polluted water [4]. Significantly, there is a substantial amount of ongoing research focused on using photocatalysis in several domains such as hydrogen production, organic matter transformation, carbon dioxide reduction, self-cleaning surface coatings, and numerous other applications. Returning to our initial concern, considerable enthusiasm exists surrounding the application of heterogeneous photocatalysis in the purification of water. This is primarily attributed to its reliance on accessible visible or solar light, its ability to mineralize organic compounds, its straightforward methodology, and its minimal equipment requirements [2].

This section provides a thorough examination of recent scholarly investigations on the process of photocatalytic destruction of organic contaminants. This study provides a comprehensive overview and analysis of the principal results and progress made in this particular area of research.

5.2 BiOX photocatalysts: principle pollutant degradation mechanism

Photocatalytic degradation encompasses the processes of absorbing light energy and generating charge carriers, as well as the migration and separation of these charge carriers to the catalyst's surface. It also involves redox reactions between the adsorbed water molecules as well as excited charge carriers. In the context of photocatalysis, the production of electron–hole pairs occurs when photons possessing an energy level that matches or exceeds the band gap of the photocatalyst are assimilated. This absorption process leads to the excitation of electrons from the valence band to the conduction band, thereby generating charge carriers. Consequently, holes are left behind in the valence band of the photocatalyst. The photons responsible for the generation of electron–hole pairs subsequently move towards the catalyst surface. The electrons and holes formed by light absorption then undergo reactions with the adsorbed oxygen and hydrogen dioxide or hydroxyl (–OH) groups present on the catalyst's surface. This leads to the formation of reactive species, including the $O_2^{\bullet-}$ as well as the $^{\bullet}OH$ radicals. Then, the reactive radical species engage in a redox interaction with the hazardous molecules, leading to the breakdown of molecules into diverse intermediates. These intermediates are then mineralized, producing CO_2, H_2O, and various inorganic anions. Furthermore, it is worth noting that the presence of generated holes in the valence band exhibits significant oxidizing capabilities. This characteristic can directly oxidize the contaminants that have been adsorbed onto the catalyst's surface. Similarly, the photoexcited electron in the conduction band possesses a significant reduction potential. Furthermore, the removal of organic contaminants can be achieved through the production of hydrogen peroxide (H_2O_2), which is formed by the

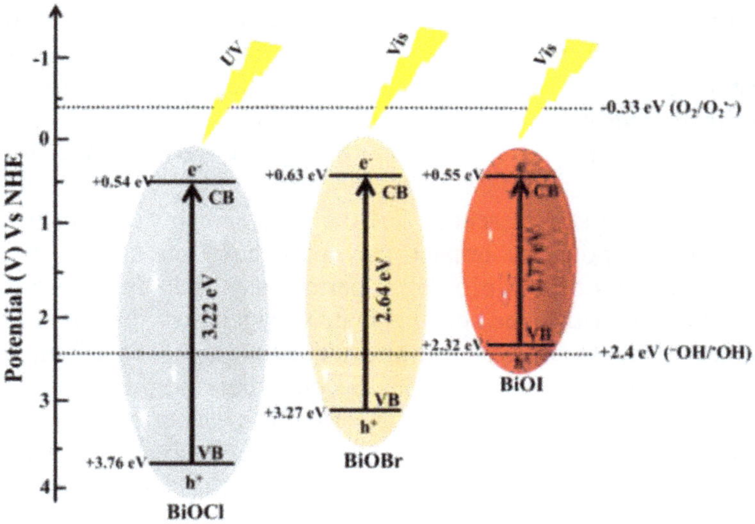

Figure 5.1. Redox potentials and band position of BiOX in photocatalysis. (Reproduced with permission from [7]. Copyright 2021 Elsevier.)

reaction between a proton (H^+) and oxygen radicals ($O_2^{\bullet-}$). It is widely acknowledged in previous research that for a photocatalytic material to exhibit exceptional performance, the redox potential or the energy levels of the valence and conduction bands must be significantly greater than the energy required for the generation of hydroxyl radicals ($O_2^{\bullet-}$ ($O_2/O_2^{\bullet-}$ = 0.33 eV/NHE) and $-OH/^{\bullet}OH$ = +2.4 eV/NHE) [5]. BiOX-based photocatalysts have the greatest photo-response when compared to other bismuth-based photocatalysts. In particular, the BiOI photocatalyst exhibits a greater visible light (VL) absorption than the BiOBr photocatalyst (2.64 eV) and the BiOCl photocatalyst (3.22 eV). The valance band potentials of BiOCl, BiOBr, and BiOI were found to be +2.32 eV, +3.27 eV, and +3.76 eV for BiOX, respectively. It was discovered that the conduction potentials of BiOI, BiOBr, and BiOCl were, in order, +0.55 eV, +0.63 eV, and +0.54 eV. It is very difficult for BiOI to achieve such a redox potential with such a tiny band gap; however, BiOBr and BiOCl can fulfill such criteria due to their greater band gaps (figure 5.1). BiOX photocatalysts, in general, have a low reducing power as a result of their generally positive conduction band position [6].

5.3 Degradation of organic pollutants

The effective treatment of bacterial illnesses using antibiotics has made them one of microbiology's greatest triumphs. In recent years, people's reliance on antibiotics has increased and their use of these drugs has increased to unhealthy levels. Based on statistical data, the global use of antibiotics in 2015 amounted to 35 billion specified daily doses [8]. Unfortunately, the human body and the bodies of animals are not capable of entirely metabolizing the antibiotics that are taken; instead, these antibiotics are expelled in part through the feces and the urine. As a consequence, parent chemicals and their metabolites can be found in high concentrations in the wastewater of municipal facilities, as well as in surface water and groundwater. In 2013, an estimated quantity of 53 800 tons of antibiotics was discharged into China's rivers and other water bodies [9]. Although antibiotics have a shorter half-life than persistent organic compounds [10], their prolonged usage has made them pseudo-permanent [11].

Antibiotics that are still present in aquatic environments are raising increasing concerns about the potential impact they may have on human health. First, exposure to antibiotics over a prolonged period may alter the distribution of the microbial population within the aqueous matrix, which may then have an effect on other complex living species via transmission down the food chain, putting the integrity of the ecosystem in jeopardy [12]. Second, the widespread use of antibiotics in low concentrations can encourage the selection of genes and bacteria that are resistant to antibiotics [13]. Third, antibiotic residues have been shown to pose significant dangers to human health, including acute and chronic toxicity hazards as well as serious injury. For instance, an excessive amount of tetracycline might impede the growth of children's bones [14] and sulfonamides can readily trigger allergic reactions in those already predisposed to them [15]. Therefore, developing reliable technology that can completely remove antibiotics from water is an absolute necessity. Recently, some different approaches have been utilized to find a solution to the issue of antibiotic contamination [16].

The usage of semiconductor-based photocatalysis for the removal of antibiotics represents a highly efficient and environmentally beneficial method with promising potential for the decomposition of harmful antibiotics [17]. Even under gentle operating conditions, the oxidative radicals produced from semiconductor photo-excitation eventually breakdown these antibiotics into smaller molecules or mineralize them into H_2O and CO_2 [18]. In the past few decades, a significant amount of effort has been put into the development of novel photocatalysts for the elimination of antibiotics with an adequate level of effectiveness [19]. For instance, Zhu *et al* [20] employed nanosized TiO_2 to breakdown tetracycline in water while it was exposed to UV irradiation. After 60 min, 95% of the tetracycline at 20 mg l^{-1} was removed. Nevertheless, the excitation of TiO_2 is limited to ultraviolet radiation with wavelengths below 387.5 nm, constituting a mere 4% of the entire solar light spectrum. Therefore, there is a strong imperative to improve the development and synthesis of efficient photocatalyst systems that exhibit improved photocatalytic capabilities and can be effectively stimulated by VL irradiation to optimize the utilization of the solar energy spectrum. The exceptional PCA of BiOX materials has been demonstrated due to their distinctive layered structures, which facilitate the efficient separation of photoexcited charge carriers. As the atomic number of the halogen increases, the band gap of BiOX decreases (BiOF has a band gap of 3.6 eV, while BiOCl has a band gap of 3.5 eV, BiOBr a band gap of 2.6 eV, and BiOI a band gap of 1.8–1.9 eV). BiOCl, classified as a p-type semiconductor, possesses a band gap energy of 3.5 eV and can absorb ultraviolet (UV) radiation [21]. However, the photo-corrosion stability of this material is significantly higher than that of the VL-absorbing compounds BiOI and BiOBr. Under VL, BiOI microspheres (MSs) have been effectively utilized to break

TC, and after 120 min of operation, 94% of TC was eliminated, but the comparable elimination rate of TOC was only 28.68% [22]. The findings suggested that the majority of TC was converted into intermediate products but that it was not fully mineralized. Most 3D or porous BiOI catalysts were produced in complex solvent systems at high pressure and temperature for protracted reaction times. BiOBr is known to be the most effective photocatalyst for antibiotic breakdown when activated by VL [23], making it one of the best BiOX materials. For instance, Zhang *et al* [24] utilized a hydrothermal process to produce flake-shaped BiOBr, and they found that ciprofloxacin (CIP), which had an initial concentration of 5 mg l^{-1}, was destroyed after 140 min when exposed to VL irradiation. In the context of the VL-induced photocatalytic process on BiOBr, it is noteworthy that the direct hole oxidation process primarily drives the oxidation of CIP. In contrast, the presence of the active species $^{\bullet}OH$, commonly observed in advanced oxidation processes, does not yield significant results. Furthermore, it has been observed that the removal of CIP was not fully achieved and was closely correlated with the resistance exhibited by the core quinolone ring.

To circumvent these obstacles, a significant amount of work has been put forward to optimize the reduction of antibiotic use. The advancements in this field have encompassed the introduction of metal doping and the fabrication of heterojunctions. The insertion of non-metal or metal ions through doping generally results in the creation of an intermediate energy level, which aims to reduce the width of the energy band. In addition, Lv *et al* [25] discovered that the incorporation of copper (Cu) into BiOBr has the potential to increase the adsorption capacity between the norfloxacin and photocatalyst. This capacity increase is generally regarded as the primary cause for improving efficiency. After 90 min, 99% of the norfloxacin was eliminated using Cu-doped BiOBr micro flowers. This removal rate is 2.28 times greater than the removal rate achieved by undoped BiOBr. In their study, Jiang *et al* manufactured a hollow MSs structure of BiOBr doped with iron (Fe). The experimental results revealed that the Fe doped BiOBr structure exhibited a much greater rate of rhodamine (RhB) degradation (1.01 h^{-1}) when compared with pure BiOBr (0.23 h^{-1}) in the presence of daylight lamp irradiation [26]. Liu *et al* demonstrated that introducing an Al dopant in BiOBr under the influence of VL irradiation resulted in enhanced PCA for the decolorization of methyl orange (MO), surpassing the effectiveness of pristine BiOBr [27]. The process of doping BiOX with Fe(III) is a highly successful technique for boosting the material's efficacy as a photocatalyst under VL irradiation. The enhanced photocatalytic efficiency can be assigned to the addition of Fe dopants, which make it easier for photogenerated carriers to be transferred [28]. Rameshbabu *et al* [29] synthesized ultrathin nanosheets of BiOCl that were modified using Cu(II). The inclusion of a metal dopant in the material serves as a means to introduce trapping centers, which effectively increase the rate of recombination between photoexcited hole and electron pairs. This is achieved by introducing additional energy levels below the conduction band or above the valence band, which can act as traps for both holes and electrons. On the surface of the BiOCl, Cu(II) was changed, which assisted in establishing conductive channels for photo-stimulated electrons. The presence of conductive routes made it possible for electrons to be drawn to Cu (II), inhibiting charge pairs to recombine. As a consequence, the PCA for breaking down MO was greatly improved. The catalysts exhibited a degradation effectiveness of 97% within a 50 min timeframe when the concentration of MO dye was 10 mg·l^{-1} (0.03 mM) and exposed to VL with a Cu (II) loading of 6%. In comparison, the pure BiOCl catalyst demonstrated a MO degradation efficiency of 75%. Even until the completion of the fifth iteration, the solution containing 6% Cu–BiOCl revealed encouraging performance, with just a marginal decline of 1.8% in its efficacy.

In addition to doping with metal ions, non-metals such as fluorine, boron, sulfur, and other such elements have also been used as doping entities. On the surface of the glass, nitrogen (N) and sulfur (S) co-doped BiOBr nanosheets were immobilized by Jiang *et al* [30]. The flawed bands can be rendered harmless by non-metal dopants since they slow down the recombination rate. Doping with non-metallic elements has resulted in the formation of S–BiOCl, C–BiOCl, and N–BiOCl in BiOX. The introduction of non-metals as dopants in BiOX leads to the generation of a narrow energy gap due to the integration of the occupied orbitals into the valence band. The introduction of heteroatoms through doping has demonstrated a significant capability to modify the band gap of BiOX effectively. This, in turn, enables the photoinduced charge carriers to be separated quickly by using an interfacial p–n junction via the efficient transfer of charge at the interface among the two components [31, 32]. Zhang *et al* [33] have created an F-doped BiOCl photocatalyst that has the maximum efficiency and is highly stable. The F/Bi molar ratio in this catalyst was 3:4. The presence of B governs the expansion of the (001) plane in BiOCl, which results in an increase in surface area. Phenol, bisphenol A (BPA), and RhB can all be more effectively eliminated with the ideal B-doped BiOCl [34]. Recent research has shown that a VL-driven S-doped BiOBr photocatalyst that was synthesized using a solvothermal method is an effective catalyst for the breakdown of ibuprofen [35]. It was found that the band gap is narrowed as a result of the S-doping procedure, which also makes the production of hydroxyl radicals more advantageous. In addition, it was discovered that utilizing a bimetallic combination resulted in a rather high photodegradation efficiency. Talreja *et al* evaluated VL-active BiOI photocatalysts doped with zinc (Zn) and Fe doped against the breakdown of tetracycline in water. Introducing bimetal doping in BiOI enhances the potential for light absorption, defects, the number of active sites, and a decelerated rate of photocharge recombination. It has been determined that Fe/Zn–BiOI-1-1 has a tetracycline degradation efficiency of around 98% [36]. Irradiating the Zn metal-based photocatalysts with light causes them to produce hydroxyl radicals ($^{\bullet}$OH), as a result of their ability to absorb significant quantities of ultraviolet light. In addition to this, the presence of Fe metal lowers the rate of electron and hole recombination, which results in a longer lifespan for the electron and hole separation. Similarly, it has been observed that Fe and Cu-doped BiOBr exhibited improved PCA (RhB and tetracycline) [37]. Jiang *et al* found that doping BiOBr flower-like MSs with Ag and Ti resulted in a greater PCA and good performance when exposed to the same daylight lamp illumination [38]. The efficient photo adsorption and separation of charge can be credited for the increased PCA that was observed (figure 5.2).

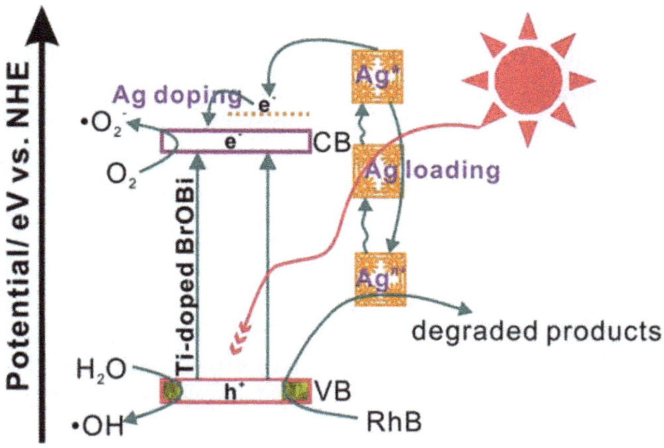

Figure 5.2. The possible reaction mechanism for decolarization of RhB over T–Ag photocatalyst. (Reproduced with permission from [38]. Copyright 2012 American Chemical Society.)

The successful development of effective nanocomposites based on BiOX for the elimination of organic contaminants frequently necessitates the synergistic interaction of many systems. Yu *et al* [39] revealed that oxygen vacancies (OVs) containing Bi/BiOCl heterojunctions displayed remarkable PCA under VL for persistent organic pollutants and dyes. The reduction of the band gap to the VL range in the BiOCl photocatalyst is principally assigned to the existence of multiple OVs on its surface. The Bi nanoparticles that are formed on the surface of the BiOCl photocatalyst enhance the efficiency of electron transfer from BiOCl to Bi in response to photoexcitation. The photodegradation process is primarily driven by photoinduced holes and $^{\bullet}O_2^-$ as the active species [39]. The process of reaction elucidates the interaction between organic contaminant molecules and active holes in the valence band of BiOCl, leading to the formation of radical ions. The aforementioned radical ions subsequently undergo reactions with $^{\bullet}O_2^-$ resulting in the production of the ultimate inorganic compounds. In addition, the composite of the I–BiOCl/I–BiOBr exhibits enhanced PCA for the breakdown of phenol and MO under VL conditions. This improvement can be attributed to the synergistic effects resulting from forming a BiOCl/BiOCl heterojunction and doping of I ion [40].

It is widely believed that the primary reactive species generated in the majority of reactions utilizing BiOX photocatalysts for the degradation of organic contaminants are $^{\bullet}O_2^-$ radicals and photoinduced holes [39, 41, 42]. According to the results of an experiment conducted by Xiao and colleagues [43], $^{\bullet}O_2^-$ radicals and photoinduced holes were discovered to play critical roles in the removal of BPA by BiOI. BiOX photocatalysts with photoinduced holes are unable to react with H_2O/OH^- to produce $^{\bullet}OH$ radicals since the typical redox potential for Bi(V)/Bi(III) is +1.59 V, which is smaller compared to the potential of $^{\bullet}OH/OH^-$ (+1.99 V) [44]. In addition, a number of investigations have shown that the two major reactive species involved in the photocatalytic process are $^{\bullet}OH$ and photoinduced holes [45]. Gao *et al* [46] developed a highly efficient three-dimensional hollow magnetic heterojunction composed of BiOI/BiOBr/Fe_3O_4. The primary objective of this heterojunction was to effectively eliminate tetrabromobisphenol (BPA) upon exposure to visible light. The experimental findings revealed that the capture of radicals in the BiOI/BiOBr/Fe_3O_4 system was mostly influenced by the presence of $^{\bullet}OH$ and photogenerated holes, which were identified as the most significant active species. $^{\bullet}OH$ was produced as a by-product of O_2 reduction by the following chain of reactions: photoinduced electrons interacted with adsorbed O_2 to give $^{\bullet}O_2^-$, which in turn produced H_2O_2; H_2O_2 then reacted with an electron to produce $^{\bullet}OH$. Using BiOCl microsheets and In_2O_3 nanoparticles, Xu *et al* [47] created a stepped heterojunction in an S-scheme configuration. The 20% In_2O_3–BiOCl exhibited increased photocatalytic degradation efficacy, resulting in a 91% removal rate of ciprofloxacin (CIP) within 35 min. The observed rate was found to be 39.6 times for pure In_2O_3 which is greater than 3.2 times for BiOCl. Additionally, using a 20% In_2O_3–BiOCl composite maintains the photodegradation efficiency of CIP at a consistent level even after undergoing five consecutive cycles. Furthermore, the assessment of chemical stability was conducted by evaluating the photodegradation efficiency of the substance under different solution pH conditions. In solutions with pH between 6 and 8, the rates of CIP degradation are greater than 91%. Jiang *et al* [48] manufactured an *in situ* manufacturing method using a pH control strategy to fabricate a Z-scheme heterostructure photocatalyst. This approach aimed to facilitate interfacial contact and effectively eliminate certain contaminants in aqueous solutions in the presence of VL conditions. Under the irradiation of VL, the Z-scheme heterostructure composed of Bi_3O_4Cl and $Bi_{12}O_{17}Cl_2$ demonstrated photocatalytic degradation activities, which degraded the pollutants that were the focus of the study. Strong interfacial coupling effects were observed on the $Bi_{12}O_{17}Cl_2$ surface due to the Z-scheme heterostructure (figure 5.3(a)). Liu *et al* successfully developed a Z-scheme 2D/2D CoAl–LDH/BiOBr photocatalyst [49]. In the Z-scheme configuration (figure 5.3(b)), it is hypothesized that the remaining vacancies in the valence band of BiOBr may undergo a reaction with H_2O, leading to the development of hydroxyl radicals. Additionally, the electrons kept in the conduction band of CoAl–LDH are expected to reduce O_2, leading to the generation of superoxide radicals. Meanwhile, an additional fraction of the electrons may induce the activation of the

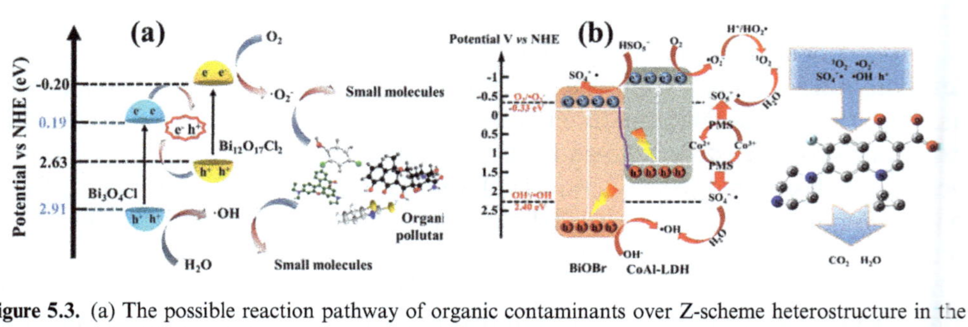

Figure 5.3. (a) The possible reaction pathway of organic contaminants over Z-scheme heterostructure in the presence of the VL irradiation. (Reproduced with permission from [48]. Copyright 2020 Elsevier.) (b) Schematic illustration for decolorizing CIP with an 8 wt% LDH/BiOBr/PMS/vis system. (Reproduced with permission from [49]. Copyright 2021 Elsevier.)

photogenerated metal–semiconductor (PMS) to produce $SO_4^{-\bullet}$, thereby facilitating the degradation of CIP as well as promoting the charge carrier's separation. Tong *et al* [50] established a direct Z-scheme heterojunction by combining phosphotungstic acid and BiOCl, denoted as BiOCl–HPW. During the experimental evaluation of photodegradation performance of TC, it was shown that the PCA of BiOCl–HPW was superior to that of both pure BiOCl and HPW. It is widely recognized that HPW exhibits a significant capability for oxidation and acidity [51]. The utilization of HPW in the system enabled the efficient separation of photoexcited charger carriers, as well as the transmission of electrons on both the interlayer and surface of BiOCl–HPWs. As a consequence, there was a notable enhancement in the quantity of photons engaged in the process of photocatalytic degradation.

Nevertheless, a significant portion of scholarly investigations has been dedicated to examining the photodegradation capabilities of BiOX, while the adsorption potential of BiOX nanoparticles has been predominantly overlooked [40, 41]. Li and coworkers [52] conducted a study wherein they synthesized BiOBr MSs to investigate the adsorption mechanism in BiOBr photocatalysis, specifically in the elimination of ibuprofen (IBP). During the process of adsorption, the rates of removal for TOC and IBP were approximately 65% and 52%, respectively. During the process of photocatalysis, there was a noticeable increase in the elimination rate, reaching values of 80% and 63% for the corresponding cases. The marginal disparity in the removal efficiency obtained between the reaction of photocatalysis and adsorption suggests that the predominant mechanism for removal is most likely the adsorption of BiOBr MSs. This finding indicates that the efficacy of removing IBP using BiOBr MSs by photocatalysis is heavily influenced by the adsorption capacity of the particles utilized. In addition, the enhanced capacity of the produced catalysts for adsorption may play a role in the effective PCA. In addition, they discovered that the key active forces involved in the IBP removal process include the production of an O–Bi–O bond as well as the ion exchange that takes place between the bromide ion and the dissociated IBP [52].

It has been determined that an efficient approach for increasing the separation effectiveness of charge as well as enhancing light-harvesting abilities is the generation of ternary nanostructures with strong contact between three different semiconductors with acceptable band locations. These nanostructures have three suitable band positions. This type of ternary structure is characterized by a powerful synergistic impact as well as interacting surface areas. Consequently, using ternary nanostructures through the combination of BiOX with diverse supplementary photocatalysts is anticipated to be a promising approach for strengthening the photoefficiency of the material. To this day, a number of ternary composites of BiOX have been described after being coupled with other materials such as carbonaceous materials, organic semiconductors, noble metal oxides, and metal oxides, amongst others. In this view, Li *et al* [53] utilized an ethylene glycol (EG) aided solvothermal approach to produce a microspherical-shaped ternary composite of Fe_3O_4/BiOBr/BiOI for RhB degradation when subjected to VL irradiation. After 80 min of illumination, the bare BiOBr and BiOI deteriorated approximately 21% and 29% of the RhB, respectively; however, the bare Fe_3O_4 exhibited practically no deterioration. It is important to point out that the PCA of the binary composite consisting of BiOBr/BiOI in a ratio of 3:1 exhibited a significant deterioration rate, amounting to around 100%. The observed increase in PCA is due to the larger surface area as well as reduced recombination of hole and electron pairs in the binary composite of BiOBr/BiOI (3:1) (figure 5.4). On the other hand, it has been observed that the elimination rate was considerably decelerated in the ternary composite of x-Fe_3O_4/BiOBr/BiOI (3:1) with varying quantities of Fe_3O_4 ($x = 0.5$, 0.4, 0.3, and 0.2). A possible cause for this phenomenon was owing to the existence of Fe_3O_4, which may partially obstruct the active sites, hence leading to insufficient PCA.

A two-step methodology that integrates ion exchange and hydrothermal methods to eliminate TC and RhB dye from aqueous solutions was used by Tang *et al* [54] to manufacture a ternary composite of AgBr-coated g-C_3N_4/BiOBr. The photodegradation efficacy of each of the AgBr/g-C_3N_4/BiOBr ternary composites is significantly higher compared to the binary composite consisting of g-C_3N_4/BiOBr and pristine BiOBr. The improved AgBr/g-C_3N_4/BiOBr-0.2 showed a breakdown of RhB of about 98% after 60 min and a degradation of TC of approximately 78% after 120 min. The improved visible light PCA may be attributed to the greater separation as well as migration of charge carriers on the surface of the AgBr/g-C_3N_4/BiOBr composite and the sluggish recombination of these pairs. Yuan *et al* [55] created a novel n–p–n heterojunction consisting of a ternary composite of AgI/BiOI/g-C_3N_4 for the breakdown of MO and TC when subjected to light irradiation from an LED. Under the irradiation of LED lights, the optimum 1-AgI/BiOI/g-C_3N_4 composite had the greatest efficiency regarding TC and MO breakdown. This increased PCA was due to the creation of an n–p–n heterojunction, which sped up the efficacy of photogenerated charge transfer and reduced the rate at which charges recombined (figure 5.5). A Z-scheme ternary composite of BiOI/Pt/g-C_3N_4 was devised by Jiang *et al* [56] to facilitate the breakdown of tetracycline hydrochloride (TCH) and phenol

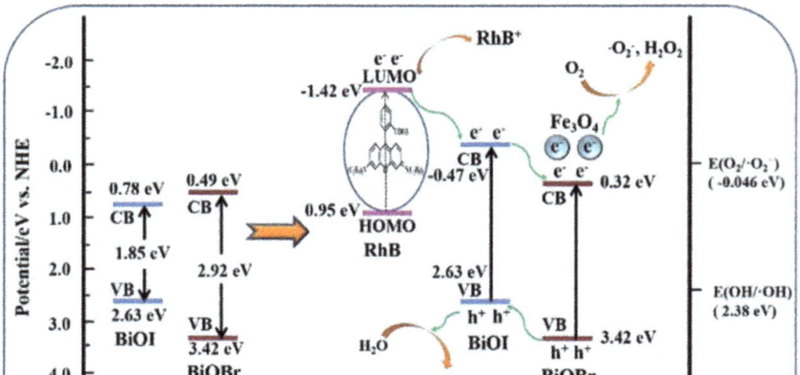

Figure 5.4. The elimination pathway for RhB removal over Fe_3O_4/BiOBr/BiOI under visible light irradiation. (Reproduced with permission from [53]. Copyright 2019 The Royal Society of Chemistry.)

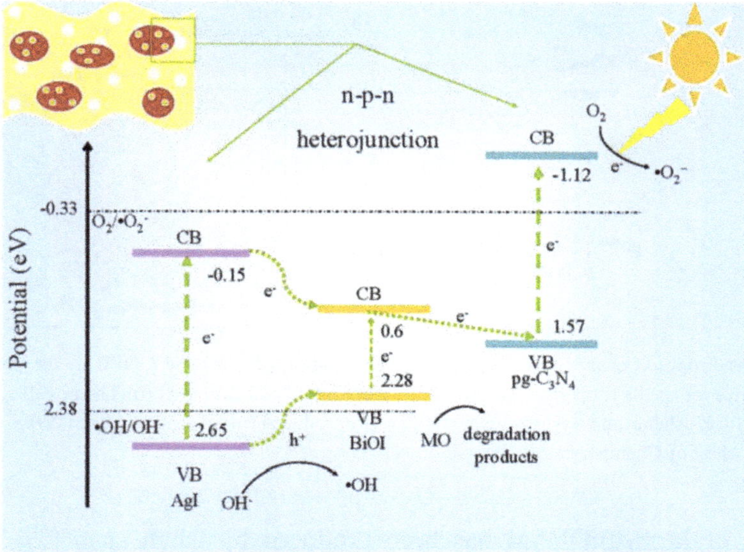

Figure 5.5. Possible degradation pathway for MO with 1-AgI/BiOI/g-C_3N_4 composite. (Reproduced with permission from [55]. Copyright 2020 Elsevier.)

in the presence of VL irradiation. The optimized ternary composite of BiOI/1%-Pt/g-C_3N_4 demonstrated the highest breakdown efficiency for eliminating TCH (83% in less than 30 min) and phenol (71% in less than 120 min). The increased degradation efficiency is made possible by the powerful capacity of Pt to capture electrons and the development of an interfacial electric field among g-C_3N_4 and BiOI. Zhong *et al* [57] used an ion exchange and two-step reflux technique PCA to create a g-C_3N_4/BiOI/$Bi_2O_2CO_3$ ternary composite, which was then tested for RhB degradation in the sunshine. It was observed that the ternary composite made up of g-C_3N_4/BiOI/$Bi_2O_2CO_3$ had a substantially greater RhB degrading efficiency than the binary composite made up of g-C_3N_4 and BiOI, as well as bare g-C_3N_4. This occurred due to the robust adsorption capacity and the enhanced separation efficiency of photo-generated charge carriers.

A ternary Z-scheme heterojunction of g-C_3N_4/PDA/BiOBr was manufactured by first varying g-C_3N_4 with polydopamine (PDA) and then depositing BiOBr using a solvothermal technique [58]. Under the light of the VL irradiation, the PCA of the ternary composite composed of g-C_3N_4/PDA/BiOBr was investigated for its ability to facilitate the breakdown of sulfamethoxazole (SMX). SMX solution degradation was ~100% for g-C_3N_4/PDA/BiOBr ternary composite compared to pristine and binary samples. The utilization of PDA as an electron transfer mediator enhanced the interfacial charge transfer and photogenerated charge separation between g-C_3N_4 and BiOBr, resulting in the improved performance of the g-C_3N_4/PDA/BiOBr system (figure 5.6(a)). The investigation on the reusability and stability of g-C_3N_4/PDA/BiOBr found that 88% of SMX was removed after five cycles of trials that proved the reusability and stability of ternary nanocomposites. A Z-scheme ternary

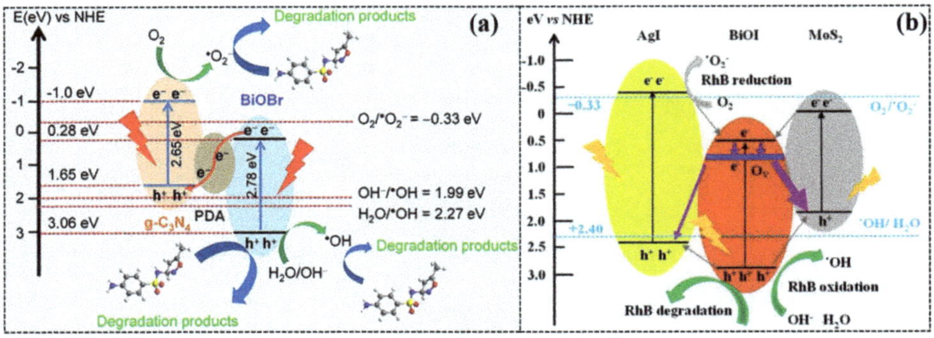

Figure 5.6. (a) Proposed charge transfer and photoelimination of SMX by CNPB in the presence of VL irradiation. (Reproduced with permission from [58]. Copyright 2020 Elsevier.) (b) The possible photocatalyst mechanism with Z-scheme and oxygen vacancy states. (Reproduced with permission from [59]. Copyright 2016 The Royal Society of Chemistry.)

composite of MoS_2/BiOI/AgI has been produced by Islam *et al* [59] utilizing a precipitation procedure for RhB degradation while it was exposed to VL irradiation. The ternary composite consisting of 2%-MoS_2/BiOI/AgI exhibited the highest RhB degrading efficiency, surpassing the effectiveness of bare BiOI by roughly 16-fold. This increased PCA of the ternary composite was attributed to the successful photoexcited charge separation using a Z-scheme mechanism, as illustrated in figure 5.6(b).

The preparation of an Ag/Ag_2O/BiOCl ternary composite was performed by Zhao *et al* [60] using a two-step calcination oxidation and photoreduction technique. The objective of the study was to facilitate the photocatalytic breakdown of RhB under VL irradiation. After a duration of 120 min under illumination, the Ag/Ag_2O/BiOCl ternary composite achieved a degradation rate of 91.2% for RhB which seems to be 32-fold higher when compared with pristine BiOCl. The effective photoactivity might be due to the efficient photoexcited charge separation and the delayed recombination of those charges. An ultrasound-aided hydrothermal approach was utilized by Zhang *et al* [61] to create an Ag/$AgVO_3$/BiOCl ternary composite for the breakdown of methylene blue (MB) under VL irradiation. The ternary composite of Ag/$AgVO_3$/BiOCl-0.5 wt% exhibited superior efficiency. Approximately 93.16% of the molecular compound MB underwent dissolution within one hour after being subjected to VL. The efficiency exhibited by the Ag/$AgVO_3$/BiOCl-0.5 wt% system was found to be higher compared to both the binary composite of Ag/$AgVO_3$ (37.52%) and the bare BiOCl (29.24%). The improved efficiency of the ternary composites Ag/$AgVO_3$/BiOCl concerning PCA can be ascribed to their greater light absorption capability and enhanced charge separation efficiency (figure 5.7). Galvan *et al* [62] created a unique ternary composite of Au-anchored BiOCl/$BiVO_4$ by performing a chemical reduction of $HAuCl_4$ in the presence of ascorbic acid. The authors then investigated the photoactivity of the composite for MO breakdown while it was exposed to VL irradiation. The optimal 1.50% Au-loaded BiOCl/$BiVO_4$ sample displayed the maximum degrading performance (67%) after 2 h of exposure

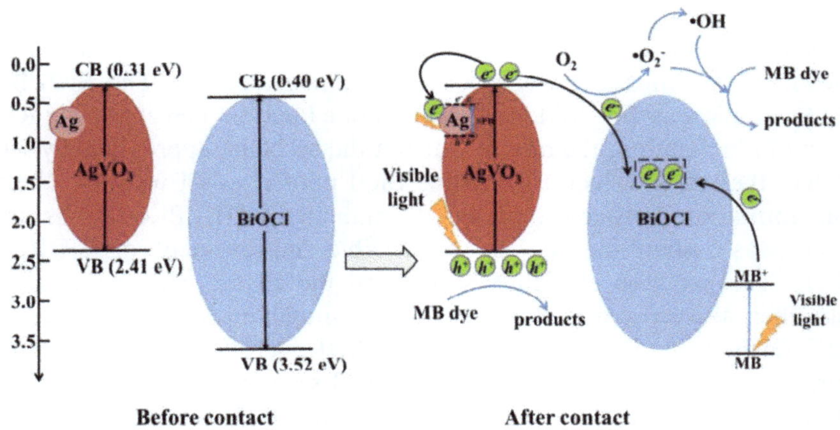

Figure 5.7. The band structures and photoreduction pathway of MB over Ag/$AgVO_3$/BiOCl photocatalyst. (Reproduced with permission from [61]. Copyright 2015 The Royal Society of Chemistry.)

to VL when compared to all of the samples. The synergistic impact of Au/BiOCl/$BiVO_4$ contributed to increased MO photoelimination. Palmai *et al* [63] utilized a one-pot preparation method to manufacture a ternary composite of Pd-decorated $BiVO_4$/BiOBr material. The PCA of the materials in their as-prepared state has been investigated for the breakdown of RhB and 4-chlorobiphenyl. The $BiVO_4$/BiOBr ternary composite loaded with optimal Pd demonstrated improved efficiency in degrading. It was determined that the metal oxide heterojunction and palladium (Pd) nanoparticles had a synergistic relationship, which was responsible for the considerable improvement in the PCA.

Zarezadeh *et al* [64] developed a p–n–p heterojunction of BiOBr/ZnO/BiOI using a straightforward refluxing technique. They subsequently examined the photodegradation efficacy of this heterojunction by MB, fuchsine, RhB, and MO dye pollutants under VL irradiation. The optimal ternary composite of BiOBr/ZnO/BiOI-20% showed a greater rate of degradation for the removal of fuchsine, MB, MO, and RhB. The greater PCA of the ternary composite consisting of BiOBr/ZnO/BiOI-20% may be ascribed to the enhanced surface area as well as the effective charge separation by the creation of p–n–p heterojunction. Using a one-pot hydrothermal technique, Yang *et al* [65] synthesized a ternary composite consisting of $CdWO_4$/BiOCl/Bi_2WO_6 with a flower-like morphology. The PCA of this composite was examined for its efficacy in the degradation of RhB in the presence of simulated solar radiation. After 60 min of irradiation, it was observed that the optimal ternary composite of $CdWO_4$/BiOCl/Bi_2WO_6 achieved a removal efficiency of over 99% for the RhB dye. In contrast, the binary composite of BiOCl/Bi_2WO_6 exhibited a lower removal efficiency, eliminating only roughly 83% of the RhB dye. The optimal CB band gap locations of the ternary composite of $CdWO_4$/BiOCl/Bi_2WO_6 were responsible for this considerable boost, as well as the increased charge carriers separation efficacy. The p–n–p heterojunction of 1D-BiOI/$Bi_4O_5I_2$/$Bi_2O_2CO_3$ was created by Peng *et al* [66] utilizing a low-temperature solution technique and making use of Bi_2O_3 nanorods as a template. This study aimed to determine the photocatalytic performance of the as-prepared heterojunction for the breakdown of RhB in the presence of solar light irradiation. The $BiOI/Bi_4O_5I_2/Bi_2O_2CO_3$ sample was able to destroy around one hundred percent of the RhB dye in 40 min of illumination, the rate at which it did so being approximately 40 times higher than the rate at which the pristine BiOI photocatalyst worked. The p–n–p junction formation between materials containing BiOI, $Bi_4O_5I_2$, as well as $Bi_2O_2CO_3$ was responsible for the remarkable improvement of $BiOI/Bi_4O_5I_2/Bi_2O_2CO_3$. This junction efficiently suppressed the charge carrier recombination using an internal electric field. Cao *et al* [67] manufactured the ternary composite Ag/AgI/BiOI by using the three-dimensional BiOI microflower as a template. The considerably improved photodegradation of MO seen in the Ag/AgI/BiOI ternary system when compared to that observed in the pristine BiOI was largely attributed to the extremely effective hole and electron separation across the closely contacted interfaces in the ternary system of Ag/AgI/BiOI. A one-step vapor diffusion technique was used by Xu *et al* [68] in order to build an AgCl/Ag/BiOCl ternary heterojunction for MO degradation under VL irradiation. Irradiation for 15 min caused the sample of AgCl/Ag/BiOCl to breakdown to the point where less than one hundred percent of the MO solution remained after the process. The development of a heterojunction between BiOCl and AgCl may efficiently promote the process of charge transfer and increase PCA. Using a hydrothermal technique, Xu *et al* [69] prepared a ternary composite of Ag/BiOBr/rGO to degrade ketoprofen (KP) under VL irradiation conditions. For the breakdown of KP, the ternary composite made of Ag/BiOBr/rGO exhibited the highest PCA compared to the binary composite made of bare BiOBr, Ag/BiOBr, and BiOBr/rGO. Effective migrating and separating of photoexcited electrons and hole pairs were hypothesized to cause better PCA (figure 5.8).

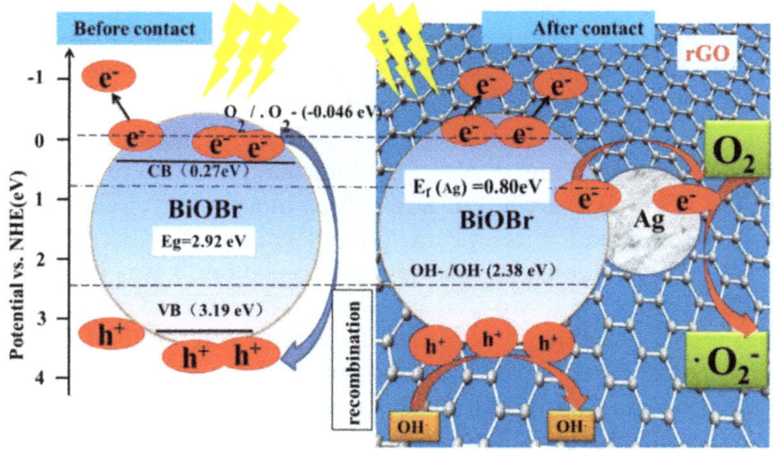

Figure 5.8. Proposed mechanism of PCA by Ag–BiOBr–rGO. (Reproduced with permission from [69]. Copyright 2019 Elsevier.)

5.4 The removal of heavy metals

It is well knowledge that certain types of heavy metals have a very high level of their respective toxins. Due to its significantly higher toxicity compared to trivalent chromium, hexavalent chromium is commonly subjected to conversion processes to transform it into the less harmful trivalent chromium form, before its removal using precipitation or adsorption mechanisms [70]. On the other hand, during this process, the reducing chemicals frequently result in secondary pollution. Heavy metals may be found in wastewater, and the sources of heavy metals and the methods used to remove and identify heavy metals are all explained in figure 5.9. Photocatalytic reduction is widely regarded as a highly effective method for removing hexavalent chromium as well as other heavy metals because it is non-toxic, has a high efficiency, and exhibits good selectivity. In the photocatalytic reduction process, heavy metal ions are adsorbed onto the photocatalysts surface, where they are then reduced. The reduced metal can then be extracted using either chemical or physical processes, depending on preference. Recent research suggests that BiOX nanoparticles offer a significant amount of potential for the photoreduction of heavy metal ions. This contrasts with the fact that BiOX nanomaterials have mostly been exploited as VL photocatalysts for the photocatalytic breakdown of organic pollutants.

Using ion exchange, Long *et al* [71] created BiOBr–Bi_2S_3 heterojunctions. The photoreduction experiments with hexavalent chromium were carried out using BiOBr–Bi_2S_3 heterostructures with the following pH values: 4, 6, 8, 10, and 12.

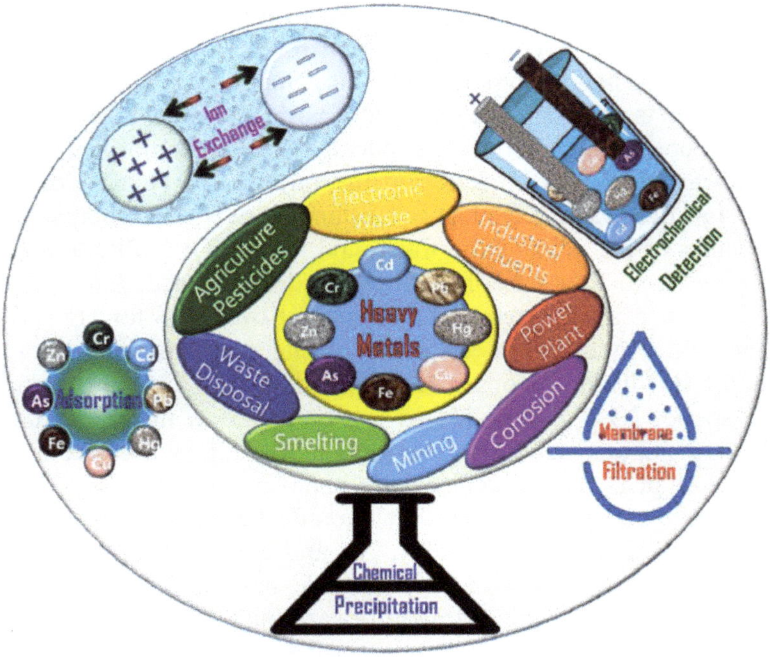

Figure 5.9. Schematic representation of heavy metals, sources of heavy metals, and techniques used to remove or detect heavy metals. (Reproduced with permission from [1]. Copyright 2022 Elsevier.)

When compared to pure BiOBr, Bi_2S_3, as well as their physical mixing, every photocatalyst had a higher hexavalent chromium removal efficiency. However, the photocatalytic efficiency was dramatically diminished when the pH was excessively high (pH 10 or above). This was because nonstoichiometric BiOBr was formed when extremely alkaline conditions were present. The BiOBr–Bi_2S_3 composite exhibited the most efficient removal of hexavalent chromium, achieving complete removal during 12 min at a pH of 6. Furthermore, the photoreduction process resulted in removal rates for hexavalent chromium that were 184.6 times and 28.9 times higher compared to pristine BiOBr and Bi_2S_3, respectively. Bai *et al* [72] used molecular precursors to generate Bi-rich $Bi_4O_5Br_xI_{2-x}$ solid solutions for the photoreduction of hexavalent chromium under VL. It may be deduced from the fact that the hexavalent chromium removal ratios of Bi_4O_5BrI, $Bi_4O_5I_2$, and $Bi_4O_5Br_2$ are, respectively, 88%, 53%, and 47%, that Bi_4O_5BrI solid solutions demonstrate a considerably greater photocatalytic reduction capacity for the elimination of hexavalent chromium than $Bi_4O_5I_2$ and $Bi_4O_5Br_2$. This effectiveness can be due to the higher conduction band position and quicker separation rate of photoexcited carriers afforded by the solid solution technique and the Bi-rich method [72]. Based on the findings of this research, a synergistic effect between bismuth-rich solid solutions and BiOX photocatalysts may be able to increase the efficiency of hexavalent chromium removal considerably. A bismuth-rich method was reported by Shang *et al* [73] to be capable of transforming BiOBr into $Bi_{24}O_{31}Br_{10}$, which had increased photocatalytic reduction activity. In the photoreduction process of hexavalent chromium under VL, $Bi_{24}O_{31}Br_{10}$ presented the maximum efficacy among BiOBr, $Bi_{24}O_{31}Br_{10}$, and Bi_2O_3, and hexavalent chromium ions were able to be reduced in 40 minutes of irradiation. $Bi_{24}O_{31}Br_{10}$ possesses a higher conduction band level when compared to BiOBr. This can be due to the hybridization of Br 4s and Bi 6p states inside the compound. Zhang *et al* [74] produced BiOX photocatalysts and examined the PCA of Hg^0 removal and the influence on SO_2 and NO while conducting their research under a fluorescent lamp. The PCA of BiOBr and BiOI toward Hg^0 was hindered by the presence of NO. In the process of eliminating Hg^0, BiOI exhibited a higher resistance to SO_2 when compared to BiOBr. The effectiveness of Hg^0 elimination was found to be BiOI > BiOBr > BiOCl. The results obtained from four experimental cycles demonstrated that both BiOI and BiOBr exhibited a consistent removal efficiency of over 80% for Hg^0. This finding suggests that BiOI and BiOBr displayed strong and sustained photoreduction capabilities for Hg^0 under the observed conditions. In the reaction system including BiOBr, the species $^{\bullet}O_2^-$ and h^+ played a crucial role in the elimination of Hg^0. Conversely, in the photocatalytic system involving BiOI, in addition to h^+, $^{\bullet}O_2^-$, and $^{\bullet}OH$, the presence of I_2 may be significant in removing higher levels of Hg^0.

The initial solution's pH has been shown to affect the photocatalytic elimination of heavy metals [75]. In a recent study, Xu and colleagues [76] produced nano-composites of BiOCl doped with boron nitride possessing a microsphere shape resembling a flower by utilizing a straightforward microwave-assisted approach for the elimination of hexavalent chromium impurities. At a pH level of 2, the 1% boron nitride-doped BiOCl demonstrates the highest capacity among all of the synthesized samples to reduce hexavalent chromium under VL photocatalytically. This ability is approximately 2.39 times greater when compared with pristine BiOCl MSs. This might be explained by the boron nitride doping, which boosts the adsorption capabilities of BiOCl, increases the material's ability to absorb VL, reduces the width of the band gap, as well as prevents the photoexcited charge carriers from recombining. Upon exposure to VL, the electrons inside the BiOCl material undergo excitation, transitioning from the valence band to the conduction band. Consequently, this process results in the creation of holes within the valence band. Following this, the photoinduced electrons in the valance band may be strapped and transmitted by boron nitride to facilitate the conversion of adsorbed hexavalent chromium to trivalent chromium [76].

The capacity of BiOX nanoparticles to adsorb heavy metal ions during the photoreduction process is not something that can be overlooked. Flower-like BiOBr nanoparticles were found to be efficient adsorbents for reducing hexavalent chromium ions throughout a broad pH range by Li's group [77]. The nanostructures observed in this study were synthesized using microwave irradiation and have exhibited a satisfactory capability for eliminating hexavalent chromium ions. The observed phenomenon can be attributed to the loose structure in the nanostructures, coupled with their substantial specific surface area. On the basis of this information, their team has constructed a series of BiOX nanostructures manufactured by a microwave-aided technique in mannitol solution [78]. The hierarchical BiOX nanostructures, like flowers, had a significantly higher capacity for removing hexavalent chromium than other BiOX nanostructures. This superior performance can be due to their unique hierarchical structures, which provide highly specialized surface areas. Among the many nanostructures investigated, the hierarchical BiOX nanostructures, resembling flowers, had the highest efficacy in removing hexavalent chromium. Compared to nanoparticles resembling flowers such as BiOBr and BiOCl, the efficacy of flower-like BiOI nanomaterials in adsorbing hexavalent chromium was significantly diminished [78]. This could be closely related to their fundamental characteristics, such as their isoelectric point (IEP), distribution of surface charge, BET surface area, etc. The IEP of flower-like BiOBr, BiOI, and BiOCl nanomaterials was calculated to be around 2.6, 0.9, and 1.9, respectively. Therefore, the slightly negative IEP of BiOI nanomaterials resembling flowers may contribute to their limited removal capacity for negatively charged hexavalent chromium species ($HCrO_4^-$ and $Cr_2O_7^{2-}$) [78].

These findings shed light on the untapped potential of BiOX photocatalytic nanomaterials in the photoreduction of heavy metal ions. Photoreduction of dangerous heavy metal ions is a technique that is both effective and economical in terms of energy consumption. In particular, the reduction performance of heavy metal ions by BiOX photocatalyst when subjected to the irradiation of VL can be significantly improved with the assistance of certain strategies, such as co-catalyst [76], crystal facet control [79–81], construction of heterojunction [82, 83], Bi-rich strategy [72, 73], and solid solutions [72]. The progress made in this study has the potential to enhance the performance of BiOX-based photocatalytic devices in the context of removing heavy metal ions.

5.5 Water disinfection

Chlorine is commonly used to disinfect water due to its effectiveness, cheap cost, and persistent impact. On the other hand, the chloro-organic by-products that are generated throughout this process have the potential to cause cancer [84]. The treatment of microbially contaminated water in hospitals and biomedical labs often involves the use of irradiation, autoclaving, detergents, or solvents, all of which have the potential to contribute to the partial removal of pathogens and, in some cases, the production of dangerous intermediates [85]. Even though some other disinfectants such as formaldehyde, hydrogen peroxide, alcohol, ozonation, UV radiations, chlorine dioxide, and so on are also used, the major demerits of these techniques include the high operational cost, expensive and complicated equipment, a lack of residual effects, generation of toxic secondary products, and so on [86]. In addition, membrane technologies are utilized, however, these membranes can only filter out bacteria and they are unable to destroy viruses. As a result, we need a method that is both cost-effective and produces non-toxic by-products to manage pathogenic micro-organisms.

5.5.1 Water disinfection mechanism by photocatalysis

Photocatalysis is a process that may eliminate micro-organisms (such as bacteria, protozoa, fungus, algae, and viruses) as well as microbial toxins (such as gram-negative endotoxins and cyanobacterial toxins). Matsunaga *et al* first claimed that the lethal effect of a photocatalyst was accomplished by demineralization, which was then followed by suppression of cell respiration brought about by the depletion of coenzyme A [87]. In addition, photocatalysts have been shown to damage the cell wall as well as the cell membrane, which ultimately results in intracellular leakage. The most frequent type of microorganism that may contaminate water is bacteria. Gram-negative and gram-positive bacteria are the two primary groups of bacteria that are responsible for a variety of death-causing infections. Gram-positive bacteria are more common than gram-negative bacteria. The outer cell membranes of these two types of bacteria are fundamentally distinct from one another [88, 89]. The exterior phospholipid layer of gram-negative bacteria, often known as the lipopolysaccharide bilayer (LPS), is asymmetrical and non-symmetrical. A peptidoglycan layer, also known as PG, and a cytoplasmic membrane, also known as a phospholipid bilayer, can be found within the LPS. Both the LPS and PG are essential components of the cell wall, which serves to defend the cell. The PG layer of gram-positive bacteria is significantly thicker than the GN layer of gram-negative bacteria. Because of its substantial PG layer composed of gram-positive bacteria, it is more resistant to the breakdown process. Other remaining phospholipids include cardiolipin (5%), phosphatidylglycerol (15%–20%), and phosphatidylethanolamine (70%–80%). However, the largest phospholipid component in gram-negative *Escherichia coli* is phosphatidylethanolamine (70%–80%) [90]. To the best of our knowledge, some of the processes that are understood are as follows: (i) metal ion release, (ii) an oxidative stress mechanism, and (iii) a non-oxidative mechanism [91]. It is still unclear what the specific mechanism is that causes cells to become inactive as a result of photocatalysis. In the process known as metal ion release, metal ions that are released from the semiconductor cause damage to the proteins and different groups of nucleic acid (–COOH, –SH, and –NH) that are contained within the microbial cell; however, this approach is not a path that leads to death. In addition, a process that does not involve oxidation can cause a reduction in the rate of cellular metabolism in bacteria, however, this mechanism is not well understood.

The oxidative stress mechanism has been the topic that has generated the most interest among researchers. During this process, a significant amount of reactive oxidative species, also known as ROS, are produced due to the redox reactions carried out by the photocatalyst [91]. Within the microbial cell, the oxidation process is driven by these ROS, which are accountable for it. When the photocatalyst is exposed to VL with an appropriate amount of energy ($hv > E_g$), the development of photogenerated electrons in the conduction band and holes in the valance band occurs. When these photoexcited charge carriers interact with dissolved O_2 molecules and H_2O, ROS such as hydrogen peroxide, oxygen radicals, and hydroxyl radicals are produced. These ROS cause damage to the cell membranes and cell walls of bacteria through their interactions with those components. ROS oxidizes the phosphatidylethanolamine in the membrane, which results in the formation of malondialdehyde [92], which is a product of lipid peroxidation. This malondialdehyde serves as a gauge for determining the level of peroxidation and damage to the cells. The penetration of catalyst nanoparticles into bacterial cells is facilitated by the electrostatic force of interaction between the photocatalyst and the bacteria, resulting in increased damage to the cell membrane. The impairment of the cell membrane and cell wall results in the release of intracellular chemicals (K^+) and amplifies the antibacterial efficacy. This K^+ is the ubiquitous chemical found in bacteria that assists in the production of proteins and the regulation of the amount of polysomes present. In addition, cell components such as ribosomes, proteins, genomic material (RNA and DNA), and so on are harmed by these ROS because of the enhanced cell permeability [93]. Figure 5.10 depicts a schematic representation of the deactivation of cells brought about by reactive oxygen species and nanoparticles.

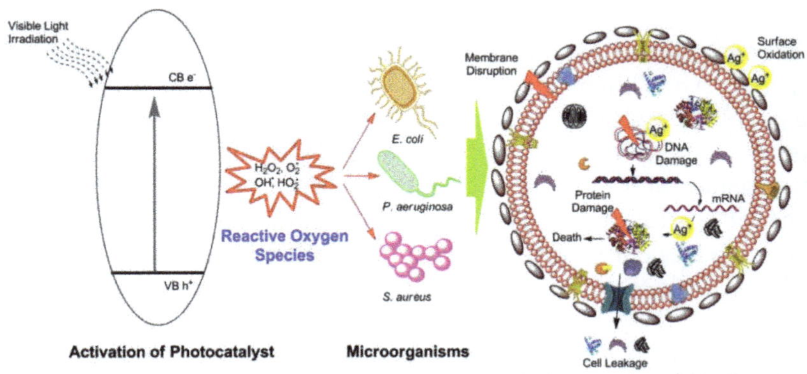

Figure 5.10. Schematic diagram of antibacterial mechanism. (Reproduced with permission from [93]. Copyright 2017 Elsevier.)

The contact of photocatalysts with micro-organisms is caused by a variety of energies, including receptor–ligand interaction, electrostatic attraction, van der Waals forces, and hydrophobic interaction. Doping, heterojunction, and morphological tuning are three techniques that might lead to the creation of a more effective photocatalyst once we have a better understanding of the mechanism by which cells are destroyed. It is more significant to study photocatalysts that are active when exposed to VL, and Bi-based photocatalysts are one such type of photocatalysts; when subjected to VL, they become activated.

5.5.2 BiOX photocatalysts for water disinfection

BiOX are non-hazardous and inert; in addition, they have excellent carrier transport qualities [94]. Studies utilizing DFT have shown that the nature of the band gap in BiOF is direct, whereas the nature in the other three is indirect. The material exhibits a favorable photoexcited charge carrier transit and separation due to the development of internal electric fields between the layers of halogen atoms and $[Bi_2O_2]^{2+}$. Consequently, it demonstrates a notable photoinduced charge carrier activity against pathogenic bacteria. Under the irradiation of fluorescent light, BiOBr MSs at a concentration of 100 mg ml^{-1}, for instance, may eliminate 90% of 1–2 × 10^7 cfu ml^{-1} of *Micrococcus lylae* in a time span of 360 min. This photocatalyst has a strong capacity for light harvesting as well as a high surface-to-volume ratio. Also, compared to BiOCl and BiOI, BiOBr has a higher level of photostability while having a lower value of photocurrent [95]. In addition, the BiOI photocatalyst demonstrated a high level of PCA against *E. coli*. Nevertheless, the photocatalyst synthesized with EG as the solvent, referred to as BiOI-EG, exhibited significantly elevated levels of PCA compared to the photocatalyst synthesized using distilled water as the solvent, denoted as BiOI-W [96]. The observed disparity in PCA can be due to the higher surface area of BiOI-EG (410 m^2 g^{-1}) compared to BiOI-W (296.6 m^2 g^{-1}), along with the effective separation of charges. Also, under the irradiation of VL, BiOI at a concentration of 1 mg ml^{-1} was able to inactivate 99.9% of bacillus sp. (3.7 106 cfu ml^{-1}) and 99.8% of *Pseudoalteromonas* sp. (2.7106 cfu ml^{-1}) within a timeframe of 60 min [97]. Attri *et al* developed nanosheets of nickel-doped BiOCl, which were analyzed for photocatalytic antibacterial activity against *Staphylococcus aureus* bacteria when illuminated by VL [98]. When *S. aureus* was exposed to light, there was a 99.5% mortality rate among the bacteria. They concluded that several active species (such as H^+, $^{\bullet}O_2^-$, and $OH^{\bullet}$ produced by the Ni–BiOCl photocatalyst) come into contact with the surface of bacteria, where they oxidize the cell wall and disrupt the cell's permeability. The process that is responsible for cell death and the loss of intracellular components, as well as biomolecules, is the one that is described above. Ultrathin nanosheets of BiOI (h-BiOI) with a thickness of approximately 2 nm were produced by Jiang and colleagues [99]. Because of the quantum size effect, the valence band of h-BiOI has a positive shift compared to the structure of block BiOI, which results in a considerable increase in the oxidation capacity. Because of this, the ability of h-BiOI to kill *E. coli* bacteria is greatly enhanced compared to block BiOIs. The strong PCA of BiOX photocatalysts may be linked to the fact that they possess a high dipole moment, one that is more than 2.0 debye. The hybridization of halogen np states decreased the hole mobility. However, it did not influence the mobility of electrons, and as a result, it improved the charge carrier separation, which eventually led to an increase in photocatalytic efficiency [100, 101].

References

[1] Perumal S, Lee W and Atchudan R 2022 A review on bismuth-based materials for the removal of organic and inorganic pollutants *Chemosphere* **306** 135521

[2] Suresh R, Rajendran S, Kumar P S, Hoang T K A and Soto-Moscoso M 2022 Halides and oxyhalides-based photocatalysts for abatement of organic water contaminants—an overview *Environ. Res.* **212** 113149

[3] Xu Z, Zhang Q, Li X and Huang X 2022 A critical review on chemical analysis of heavy metal complexes in water/wastewater and the mechanism of treatment methods *Chem. Eng. J.* **429** 131688

[4] Ong C B, Ng L Y and Mohammad A W 2018 A review of ZnO nanoparticles as solar photocatalysts: synthesis, mechanisms and applications *Renew. Sustain. Energy Rev.* **81** 536–51

[5] Jo W K and Natarajan T S 2016 Fabrication and efficient visible light photocatalytic properties of novel zinc indium sulfide ($ZnIn_2S_4$)—graphitic carbon nitride (g-C_3N_4)/bismuth vanadate ($BiVO_4$) nanorod-based ternary nanocomposites with enhanced charge separation via Z-scheme transfe *J. Colloid Interface Sci.* **482** 58–72

[6] Pare B, Sarwan B and Jonnalagadda S B 2012 The characteristics and photocatalytic activities of BiOCl as highly efficient photocatalyst *J. Mol. Struct.* **1007** 196–202

[7] Arumugam M, Natarajan T S, Saelee T, Praserthdam S, Ashokkumar M and Praserthdam P 2021 Recent developments on bismuth oxyhalides (BiOX; X = Cl, Br, I) based ternary nanocomposite photocatalysts for environmental applications *Chemosphere* **282** 131054

[8] Antoñanzas F and Goossens H 2019 The economics of antibiotic resistance: a claim for personalised treatments *Eur. J. Heal. Econ.* **20** 483–5

[9] Zhang Q Q, Ying G G, Pan C G, Liu Y S and Zhao J L 2015 Comprehensive evaluation of antibiotics emission and fate in the river basins of China: source analysis, multimedia modeling, and linkage to bacterial resistance *Environ. Sci. Technol.* **49** 6772–82

[10] Trojanowicz M 2020 Removal of persistent organic pollutants (POPs) from waters and wastewaters by the use of ionizing radiation *Sci. Total Environ.* **718** 134425

[11] Patel M, Kumar R, Kishor K, Mlsna T, Pittman C U and Mohan D 2019 Pharmaceuticals of emerging concern in aquatic systems: chemistry, occurrence, effects, and removal methods *Chem. Rev.* **119** 3510–673

[12] Meredith H R, Srimani J K, Lee A J, Lopatkin A J and You L 2015 Collective antibiotic tolerance: mechanisms, dynamics and intervention *Nat. Chem. Biol.* **11** 182–8

[13] Qin K, Wei L, Li J, Lai B, Zhu F, Yu H, Zhao Q and Wang K 2020 A review of ARGs in WWTPs: sources, stressors and elimination *Chinese Chem. Lett.* **31** 2603–13

[14] Daghrir R and Drogui P 2013 Tetracycline antibiotics in the environment: a review *Environ. Chem. Lett.* **11** 209–27

[15] Giles A, Foushee J, Lantz E and Gumina G 2019 Sulfonamide allergies *Pharmacy* **7** 132

[16] Xia X, Zhu F, Li J, Yang H, Wei L, Li Q, Jiang J, Zhang G and Zhao Q 2020 A review study on sulfate-radical-based advanced oxidation processes for domestic/industrial wastewater treatment: degradation, efficiency, and mechanism *Front. Chem.* **8** 592056

[17] Ajiboye T O, Oyewo O A and Onwudiwe D C 2021 The performance of bismuth-based compounds in photocatalytic applications *Surfaces and Interfaces* **23** 100927

[18] Sturini M, Speltini A, Maraschi F, Profumo A, Pretali L, Irastorza E A, Fasani E and Albini A 2012 Photolytic and photocatalytic degradation of fluoroquinolones in untreated river water under natural sunlight *Appl. Catal.* B **119–20** 32–9

[19] Wang H, Zhang J, Wang P, Yin L, Tian Y and Li J 2020 Bifunctional copper modified graphitic carbon nitride catalysts for efficient tetracycline removal: synergy of adsorption and photocatalytic degradation *Chinese Chem. Lett.* **31** 2789–94

[20] Zhu X D, Wang Y J, Sun R J and Zhou D M 2013 Photocatalytic degradation of tetracycline in aqueous solution by nanosized TiO_2 *Chemosphere* **92** 925–32

[21] Zhang L, Wang W, Jiang D, Gao E and Sun S 2015 Photoreduction of CO_2 on BiOCl nanoplates with the assistance of photoinduced oxygen vacancies *Nano Res.* **8** 821–31

[22] Hao R, Xiao X, Zuo X, Nan J and Zhang W 2012 Efficient adsorption and visible-light photocatalytic degradation of tetracycline hydrochloride using mesoporous BiOI microspheres *J. Hazard. Mater.* **209–10** 137–45

[23] Huo Y, Zhang J, Miao M and Jin Y 2012 Solvothermal synthesis of flower-like BiOBr microspheres with highly visible-light photocatalytic performances *Appl. Catal.* B **111–2** 334–41

[24] Zhang X X, Li R, Jia M, Wang S, Huang Y and Chen C 2015 Degradation of ciprofloxacin in aqueous bismuth oxybromide (BiOBr) suspensions under visible light irradiation: a direct hole oxidation pathway *Chem. Eng. J.* **274** 290–7

[25] Lv X, Yan D Y S, Lam F L Y, Ng Y H, Yin S and An A K 2020 Solvothermal synthesis of copper-doped BiOBr microflowers with enhanced adsorption and visible-light driven photocatalytic degradation of norfloxacin *Chem. Eng. J.* **401** 126012

[26] Jiang G *et al* 2013 Photocatalytic properties of hierarchical structures based on Fe-doped BiOBr hollow microspheres *J. Mater. Chem.* A **1** 2406–10

[27] Liu Z, Wu B, Zhao Y, Niu J and Zhu Y 2014 Solvothermal synthesis and photocatalytic activity of Al-doped BiOBr microspheres *Ceram. Int.* **40** 5597–603

[28] Xiao X and De Zhang W 2010 Facile synthesis of nanostructured BiOI microspheres with high visible light-induced photocatalytic activity *J. Mater. Chem.* **20** 5866–70

[29] Rameshbabu R, Sandhiya M, Pecchi G and Sathish M 2020 Effective coupling of Cu(II) with BiOCl nanosheets for high performance electrochemical supercapacitor and enhanced photocatalytic applications *Appl. Surf. Sci.* **521** 146362

[30] Jiang G, Li X, Wei Z, Wang X, Jiang T, Du X and Chen W 2014 Immobilization of N, S-codoped BiOBr on glass fibers for photocatalytic degradation of rhodamine B *Powder Technol.* **261** 170–5

[31] Tian F, Zhao H, Dai Z, Cheng G and Chen R 2016 Mediation of valence band maximum of BiOI by Cl incorporation for improved oxidation power in photocatalysis *Ind. Eng. Chem. Res.* **55** 4969–78

[32] Tian J, Chen Z, Jing J, Feng C, Sun M and Li W 2020 Enhanced photocatalytic activity of BiOCl with regulated morphology and band structure through controlling the adding amount of HCl *Mater. Lett.* **272** 127860

[33] Zhang S, Wang D and Song L 2016 A novel F-doped BiOCl photocatalyst with enhanced photocatalytic performance *Mater. Chem. Phys.* **173** 298–308

[34] Yu C, He H, Fan Q, Xie W, Liu Z and Ji H 2019 Novel B-doped BiOCl nanosheets with exposed (001) facets and photocatalytic mechanism of enhanced degradation efficiency for organic pollutants *Sci. Total Environ.* **694** 133727

[35] Liu Y, Hu Z and Yu J C 2021 Photocatalytic degradation of ibuprofen on S-doped BiOBr *Chemosphere* **278** 130376

[36] Talreja N, Afreen S, Ashfaq M, Chauhan D, Mera A C, Rodríguez C A and Mangalaraja R V 2021 Bimetal (Fe/Zn) doped BiOI photocatalyst: an effective photodegradation of tetracycline and bacteria *Chemosphere* **280**

[37] Qu J, Du Y, Ji P, Li Z, Jiang N, Sun X, Xue L, Li H and Sun G 2021 Fe, Cu co-doped BiOBr with improved photocatalytic ability of pollutants degradation *J. Alloys Compd.* **881** 160391

[38] Jiang G, Wang R, Wang X, Xi X, Hu R, Zhou Y, Wang S, Wang T and Chen W 2012 Novel highly active visible-light-induced photocatalysts based on BiOBr with Ti doping and Ag decorating *ACS Appl. Mater. Interfaces* **4** 4440–4

[39] Yu Y, Cao C, Liu H, Li P, Wei F, Jiang Y and Song W 2014 A Bi/BiOCl heterojunction photocatalyst with enhanced electron–hole separation and excellent visible light photodegrading activity *J. Mater. Chem.* A **2** 1677–81

[40] Jia X, Cao J, Lin H, Zhang M, Guo X and Chen S 2017 Transforming type-I to type-II heterostructure photocatalyst via energy band engineering: a case study of I–BiOCl/I–BiOBr *Appl. Catal.* B **204** 505–14

[41] Di J, Xia J, Ji M, Wang B, Yin S, Zhang Q, Chen Z and Li H 2016 Advanced photocatalytic performance of graphene-like BN modified BiOBr flower-like materials for the removal of pollutants and mechanism insight *Appl. Catal.* B **183** 254–62

[42] Chen F *et al* 2016 Enhanced photocatalytic degradation of tetracycline by AgI/BiVO4 heterojunction under visible-light irradiation: mineralization efficiency and mechanism *ACS Appl. Mater. Interfaces* **8** 32887–900

[43] Xiao X, Tu S, Lu M, Zhong H, Zheng C, Zuo X and Nan J 2016 Discussion on the reaction mechanism of the photocatalytic degradation of organic contaminants from a viewpoint of semiconductor photo-induced electrocatalysis *Appl. Catal.* B **198** 124–32

[44] Fu H, Pan C, Yao W and Zhu Y 2005 Visible-light-induced degradation of rhodamine B by nanosized Bi_2WO_6 *J. Phys. Chem.* B **109** 22432–9

[45] Xiao Y, Song X, Liu Z, Li R, Zhao X and Huang Y 2017 Photocatalytic removal of cefazolin using Ag_3PO_4/BiOBr under visible light and optimization of parameters by response surface methodology *J. Ind. Eng. Chem.* **45** 248–56

[46] Gao S, Guo C, Hou S, Wan L, Wang Q, Lv J, Zhang Y, Gao J, Meng W and Xu J 2017 Photocatalytic removal of tetrabromobisphenol A by magnetically separable flower-like BiOBr/BiOI/Fe_3O_4 hybrid nanocomposites under visible-light irradiation *J. Hazard. Mater.* **331** 1–12

[47] Xu K, Shen J, Zhang S, Xu D and Chen X 2022 Efficient interfacial charge transfer of BiOCl–In_2O_3 step-scheme heterojunction for boosted photocatalytic degradation of ciprofloxacin *J. Mater. Sci. Technol.* **121** 236–44

[48] Jiang E, Song N, Zhang X, Yang L, Liu C and Dong H 2020 *In-situ* fabrication of Z-scheme $Bi_3O_4Cl/Bi_{12}O_{17}Cl_2$ heterostructure by facile pH control strategy to boost removal of various pollutants in water *Chem. Eng. J.* **388** 123483

[49] Liu C, Mao S, Shi M, Wang F, Xia M, Chen Q and Ju X 2021 Peroxymonosulfate activation through 2D/2D Z-scheme CoAl–LDH/BiOBr photocatalyst under visible light for ciprofloxacin degradation *J. Hazard. Mater.* **420** 126613

[50] Tong H, Shi B and Zhao S 2020 Facile synthesis of a direct Z-scheme BiOCl-phosphotungstic acid heterojunction for the improved photodegradation of tetracycline *RSC Adv.* **10** 17369–76

[51] Wang S S and Yang G Y 2015 Recent advances in polyoxometalate-catalyzed reactions *Chem. Rev.* **115** 4893–962

[52] Li J, Sun S, Qian C, He L, Chen K K, Zhang T, Chen Z and Ye M 2016 The role of adsorption in photocatalytic degradation of ibuprofen under visible light irradiation by BiOBr microspheres *Chem. Eng. J.* **297** 139–47

[53] Li J, Yang F, Zhou Q, Wu L, Li W, Ren R and Lv Y 2019 Visible-light photocatalytic performance, recovery and degradation mechanism of ternary magnetic Fe_3O_4/BiOBr/BiOI composite *RSC Adv.* **9** 23545–53

[54] Tang G, Zhang F, Huo P, Zulfiqarc S, Xu J, Yan Y and Tang H 2019 Constructing novel visible-light-driven ternary photocatalyst of AgBr nanoparticles decorated 2D/2D heterojunction of g-C_3N_4/BiOBr nanosheets with remarkably enhanced photocatalytic activity for water-treatment *Ceram. Int.* **45** 19197–205

[55] Yuan D, Huang L, Li Y, Wang H, Xu X, Wang C and Yang L 2020 A novel AgI/BiOI/pg-C_3N_4 composite with enhanced photocatalytic activity for removing methylene orange tetracycline and *E. coli Dye. Pigment.* **177** 108253

[56] Jiang J, Song Y, Wang X, Li T, Li M, Lin Y, Xie T and Dong S 2020 Enhancing aqueous pollutant photodegradation via a Fermi level matched Z-scheme BiOI/Pt/g-C_3N_4 photocatalyst: unobstructed photogenerated charge behavior and degradation pathway exploration *Catal. Sci. Technol.* **10** 3324–33

[57] Zhong S, Zhou H, Shen M, Yao Y and Gao Q 2021 Rationally designed a g-C_3N_4/BiOI/$Bi_2O_2CO_3$ composite with promoted photocatalytic activity *J. Alloys Compd.* **853** 157307

[58] Guo F, Chen J, Zhao J, Chen Z, Xia D, Zhan Z and Wang Q 2020 Z-scheme heterojunction g-C_3N_4@PDA/BiOBr with biomimetic polydopamine as electron transfer mediators for enhanced visible-light driven degradation of sulfamethoxazole *Chem. Eng. J.* **386** 124014

[59] Jahurul Islam M, Amaranatha Reddy D, Han N S, Choi J, Song J K and Kim T K 2016 An oxygen-vacancy rich 3D novel hierarchical MoS_2/BiOI/AgI ternary nanocomposite: enhanced photocatalytic activity through photogenerated electron shuttling in a Z-scheme manner *Phys. Chem. Chem. Phys.* **18** 24984–93

[60] Zhao H, Liu X, Dong Y, Li H, Song R, Xia Y and Wang H 2019 A novel visible-light-driven ternary Ag@Ag_2O/BiOCl Z-scheme photocatalyst with enhanced removal efficiency of RhB *New J. Chem.* **43** 13929–37

[61] Zhang L, Yuan X, Wang H, Chen X, Wu Z, Liu Y, Gu S, Jiang Q and Zeng G 2015 Facile preparation of an Ag/$AgVO_3$/BiOCl composite and its enhanced photocatalytic behavior for methylene blue degradation *RSC Adv.* **5** 98184–93

[62] de la Garza-Galván M, Zambrano-Robledo P, Vazquez-Arenas J, Romero-Ibarra I, Ostos C, Peral J and García-Pérez U M 2019 *In situ* synthesis of Au-decorated BiOCl/$BiVO_4$ hybrid ternary system with enhanced visible-light photocatalytic behavior *Appl. Surf. Sci.* **487** 743–54

[63] Pálmai M, Zahran E M, Angaramo S, Bálint S, Pászti Z, Knecht M R and Bachas L G 2017 Pd-decorated m-$BiVO_4$/BiOBr ternary composite with dual heterojunction for enhanced photocatalytic activity *J. Mater. Chem.* A **5** 529–34

[64] Zarezadeh S, Habibi-Yangjeh A, Mousavi M and Ghosh S 2020 Synthesis of novel p–n–p BiOBr/ZnO/BiOI heterostructures and their efficient photocatalytic performances in removals of dye pollutants under visible light *J. Photochem. Photobiol.* A **389** 112247
[65] Yang C, Gao G, Zhang J, Fan R, Liu D, Zhang Y, Liu R, Guo Z and Gan S 2018 Controlled formation of a flower-like $CdWO_4$–BiOCl–Bi_2WO_6 ternary hybrid photocatalyst with enhanced photocatalytic activity through one-pot hydrothermal reaction *New J. Chem.* **42** 9236–43
[66] Peng Y, Yu P P, Zhou H Y and Xu A W 2015 Synthesis of BiOI/$Bi_4O_5I_2$/$Bi_2O_2CO_3$ p–n–p heterojunctions with superior photocatalytic activities *New J. Chem.* **39** 8321–8
[67] Cao J, Zhao Y, Lin H, Xu B and Chen S 2013 Facile synthesis of novel Ag/AgI/BiOI composites with highly enhanced visible light photocatalytic performances *J. Solid State Chem.* **206** 38–44
[68] Xu Z and Lin S Y 2016 Construction of AgCl/Ag/BiOCl with a concave-rhombicuboctahedron core–shell hierarchitecture and enhanced photocatalytic activity *RSC Adv.* **6** 84738–47
[69] Xu G, Li M, Wang Y, Zheng N, Yang L, Yu H and Yu Y 2019 A novel Ag–BiOBr–rGO photocatalyst for enhanced ketoprofen degradation: kinetics and mechanisms *Sci. Total Environ.* **678** 173–80
[70] Deng F, Luo Y, Li H, Xia B, Luo X, Luo S and Dionysiou D D 2020 Efficient toxicity elimination of aqueous Cr(VI) by positively-charged $BiOClxI_{1-x}$, $BiOBr_xI_{1-x}$ and $BiOCl_xBr_{1-x}$ solid solution with internal hole-scavenging capacity via the synergy of adsorption and photocatalytic reduction *J. Hazard. Mater.* **383** 121127
[71] Long Z, Zhang G, Du H, Zhu J and Li J 2021 Preparation and application of BiOBr–Bi_2S_3 heterojunctions for efficient photocatalytic removal of Cr(VI) *J. Hazard. Mater.* **407**
[72] Bai Y, Ye L, Chen T, Wang P, Wang L, Shi X and Wong P K 2017 Synthesis of hierarchical bismuth-rich $Bi_4O_5Br_xI_{2-x}$ solid solutions for enhanced photocatalytic activities of CO_2 conversion and Cr(VI) reduction under visible light *Appl. Catal.* B **203** 633–40
[73] Shang J, Hao W, Lv X, Wang T, Wang X, Du Y, Dou S, Xie T, Wang D and Wang J 2014 Bismuth oxybromide with reasonable photocatalytic reduction activity under visible light *ACS Catal.* **4** 954–61
[74] Zhang A, Xing W, Zhang D, Wang H, Chen G and Xiang J 2016 A novel low-cost method for Hg^0 removal from flue gas by visible-light-driven BiOX (X = Cl, Br, I) photocatalysts *Catal. Commun.* **87** 57–61
[75] Deng Y, Tang L, Zeng G, Zhu Z, Yan M, Zhou Y, Wang J, Liu Y and Wang J 2017 Insight into highly efficient simultaneous photocatalytic removal of Cr(VI) and 2,4-diclorophenol under visible light irradiation by phosphorus doped porous ultrathin g-C_3N_4 nanosheets from aqueous media: performance and reaction mechanism *Appl. Catal.* B **203** 343–54
[76] Xu H, Wu Z, Ding M and Gao X 2017 Microwave-assisted synthesis of flower-like BN/BiOCl composites for photocatalytic Cr(VI) reduction upon visible-light irradiation *Mater. Des.* **114** 129–38
[77] Li G, Qin F, Yang H, Lu Z, Sun H and Chen R 2012 Facile microwave synthesis of 3D flowerlike BiOBr nanostructures and their excellent CrVI removal capacity *Eur. J. Inorg. Chem.* **2012** 2508–13
[78] Li G, Qin F, Wang R, Xiao S, Sun H and Chen R 2013 BiOX (X = Cl, Br, I) nanostructures: mannitol-mediated microwave synthesis, visible light photocatalytic performance, and Cr(VI) removal capacity *J. Colloid Interface Sci.* **409** 43–51

[79] Li H and Zhang L 2014 Oxygen vacancy induced selective silver deposition on the {001} facets of BiOCl single-crystalline nanosheets for enhanced Cr(VI) and sodium pentachlorophenate removal under visible light *Nanoscale* **6** 7805–10

[80] Fan Z, Zhao Y, Zhai W, Qiu L, Li H and Hoffmann M R 2016 Facet-dependent performance of BiOBr for photocatalytic reduction of Cr(VI) *RSC Adv.* **6** 2028–31

[81] Han J, Zhu G, Hojamberdiev M, Peng J, Zhang X, Liu Y, Ge B and Liu P 2015 Rapid adsorption and photocatalytic activity for Rhodamine B and Cr(VI) by ultrathin BiOI nanosheets with highly exposed {001} facets *New J. Chem.* **39** 1874–82

[82] Wang Q, Shi X, Liu E, Crittenden J C, Ma X, Zhang Y and Cong Y 2016 Facile synthesis of AgI/BiOI–Bi_2O_3 multi-heterojunctions with high visible light activity for Cr(VI) reduction *J. Hazard. Mater.* **317** 8–16

[83] Wang K, Shao C, Li X, Zhang X, Lu N, Miao F and Liu Y 2015 Hierarchical heterostructures of p-type BiOCl nanosheets on electrospun n-type TiO_2 nanofibers with enhanced photocatalytic activity *Catal. Commun.* **67** 6–10

[84] Freuze I, Brosillon S, Laplanche A, Tozza D and Cavard J 2005 Effect of chlorination on the formation of odorous disinfection by-products *Water Res.* **39** 2636–42

[85] Regmi C, Kshetri Y K, Kim T H, Pandey R P, Ray S K and Lee S W 2017 Fabrication of Ni-doped $BiVO_4$ semiconductors with enhanced visible-light photocatalytic performances for wastewater treatment *Appl. Surf. Sci.* **413** 253–65

[86] Huang W J, Fang G C and Wang C C 2005 The determination and fate of disinfection by-products from ozonation of polluted raw water *Sci. Total Environ.* **345** 261–72

[87] Matsunaga T, Tomoda R, Nakajima T and Wake H 1985 Photoelectrochemical sterilization of microbial cells by semiconductor powders *FEMS Microbiol. Lett.* **29** 211–4

[88] Dalrymple O K, Stefanakos E, Trotz M A and Goswami D Y 2010 A review of the mechanisms and modeling of photocatalytic disinfection *Appl. Catal. B Environ.* **98** 27–38

[89] Coleman J P and Smith C J 2014 Structure and composition of microbes *Reference Module in Biomedical Sciences* (Amsterdam: Elsevier)

[90] Kiwi J and Nadtochenko V 2005 Evidence for the mechanism of photocatalytic degradation of the bacterial wall membrane at the TiO_2 interface by ATR-FTIR and laser kinetic spectroscopy *Langmuir* **21** 4631–41

[91] Regmi C, Joshi B, Ray S K, Gyawali G and Pandey R P 2018 Understanding mechanism of photocatalytic microbial decontamination of environmental wastewater *Front. Chem.* **6** 33

[92] Hu C, Guo J, Qu J and Hu X 2007 Photocatalytic degradation of pathogenic bacteria with AgI/TiO_2 under visible light irradiation *Langmuir* **23** 4982–7

[93] Ray S K, Dhakal D, Pandey R P and Lee S W 2017 Ag–$BaMoO_4$: Er^{3+}/Yb^{3+} photocatalyst for antibacterial application *Mater. Sci. Eng.* C **78** 1164–71

[94] Raizda P, Gautam S, Priya B and Singh P 2016 Preparation and photocatalytic activity of hydroxyapatite supported BiOCl nanocomposite for oxytetracyline removal *Adv. Mater. Lett.* **7** 312–8

[95] Bhachu D S *et al* 2016 Bismuth oxyhalides: synthesis, structure and photoelectrochemical activity *Chem. Sci.* **7** 4832–41

[96] Jamil T S, Mansor E S and Azab El-Liethy M 2015 Photocatalytic inactivation of *E. coli* using nano-size bismuth oxyiodide photocatalysts under visible light *J. Environ. Chem. Eng.* **3** 2463–71

[97] Long Y, Wang Y, Zhang D, Ju P and Sun Y 2016 Facile synthesis of BiOI in hierarchical nanostructure preparation and its photocatalytic application to organic dye removal and biocidal effect of bacteria *J. Colloid Interface Sci.* **481** 47–56

[98] Attri P, Garg P, Chauhan M, Singh R, Sharma R K, Kumar S, Lim D K and Chaudhary G R 2023 Metal doped BiOCl nano-architectures (M–BiOCl, M = Ni, Mo, Cd, Co) for efficient visible light photocatalytic and antibacterial behaviour *J. Environ. Chem. Eng.* **11** 109498

[99] Jiang Z, Liang X, Liu Y, Jing T, Wang Z, Zhang X, Qin X, Dai Y and Huang B 2017 Enhancing visible light photocatalytic degradation performance and bactericidal activity of BiOI via ultrathin-layer structure *Appl. Catal.* B **211** 252–7

[100] Chen F, Ma Z, Ye L, Ma T, Zhang T, Zhang Y and Huang H 2020 Macroscopic spontaneous polarization and surface oxygen vacancies collaboratively boosting CO_2 photoreduction on $BiOIO_3$ single crystals *Adv. Mater.* **32** 1908350

[101] Chen F, Huang H, Guo L, Zhang Y and Ma T 2019 The role of polarization in photocatalysis *Angew. Chem., Int. Ed.* **58** 10061–73

Chapter 6

Future prospects

BiOX exhibits considerable potential as a versatile and emerging photocatalytic material responsive to visible light. This property renders such materials suitable for the degradation of organic contaminants in aquatic environments. The efficacy of BiOX materials can be attributed to their layered structure, distinctive structural and electrical properties, as well as their band structure. Currently, the most often employed techniques for the synthesis of BiOX involve heat treatments, specifically solvothermal and hydrothermal procedures. These pathways enable researchers to manipulate the shape, precise surface area, dimensions, and pore volume and size. However, these procedures are both time-consuming and require a significant amount of energy. Future investigations should prioritize the development of an efficient technique for the widespread production of very appealing BiOBr photocatalysts on a significant scale. The utilization of microwave-assisted techniques enables the expeditious synthesis of BiOX. However, it is crucial to acknowledge that the examination of this approach is currently in its early stages.

In order to enhance the performance of BiOX-based semiconductor photocatalysts, investigators are employing various strategies, including heterojunction engineering, ion doping, carbon material interfacing, noble metal coupling, and facet control. Despite significant efforts being devoted to the regulation of materials based on BiOX as well as the optimization of their photocatalytic effectiveness, research in this field continues to encounter numerous problems while also presenting good prospects. Considering the perspective of 2D materials, the decrease in atomic layer thickness has the potential to introduce novel features that are absent in bulk materials. The inherent layered architecture of BiOX facilitates the facile formation of a monolayer or a few-layer configuration. Hence, it is imperative for researchers to focus on effectively controlling the BiOX synthesis at atomic-level thickness while investigating its photocatalytic capabilities. The manipulation of defects has been demonstrated to effectively modulate the photocatalytic performance of BiOX. However, it is worth noting that the oxygen vacancy defect is still the

doi:10.1088/978-0-7503-5934-4ch6

most common type of defect observed in BiOX. Additional defects, such as halogen vacancies and dual vacancies, have yet to be fabricated in BiOX. During the photocatalytic process, it is anticipated that these defects will engage in interactions with surface adsorbates. Moreover, the existence of such imperfections can potentially give rise to novel electronic states, thereby influencing the electron transfer process. Hence, the examination of the correlation across various defect types as well as their impact on photocatalysis holds significant potential for future research endeavors. Significant advancements have been made in the realm of BiOX photocatalysts thus far. However, several challenging issues remain to be addressed in this field. These include the controllable synthesis of BiOX, the identification of the most effective modification approach, and the exploration of the intricate link between microstructure and photocatalytic activity. The aforementioned issues present opportunities for further investigation, which can be categorized as follows.

The first is enhancing synthesis methodologies to align with practical requirements. Currently, the production of BiOX photocatalysts remains limited to laboratory settings, and the synthesis process typically entails multiple sequential reactions that yield modest quantities and require extended periods of time for completion. Hence, there is a pressing need to devise a straightforward, effective, and extensive approach that may be readily implemented in commercial settings. The objective of such study is to explore innovative synthesis approaches for the production of BiOX photocatalysts with distinct microstructural characteristics. The morphologies that are favored in this context encompass hierarchical flowers/spheres characterized by a substantial specific surface area, ultrathin nanosheets exhibiting prominently exposed {001} facets, and hollow nanostructures. Furthermore, it is imperative to consider the costs associated with raw materials, the duration required for preparation, and the process of mass production.

The synthesis of BiOX photocatalysts with various crystal facets, including the commonly seen {001} and {010} facets, is crucial. However, it is equally important to develop methods for synthesizing BiOX photocatalysts that expose other high-performance crystal facets. Furthermore, the variation in crystal facets gives rise to distinct free radicals generated by these facets. Consequently, it is imperative to conduct deeper investigations into the reaction mechanism between different facets in order to identify the most ideal crystal facet.

The intensity of interfacial electric fields (IEFs) resulting from the distinctive layered structure is a noteworthy attribute of BiOX photocatalysts. However, there is currently a dearth of research investigating the impact of IEF on the photocatalytic activity of BiOX. Future investigations should prioritize the in-depth examination of the precise mechanism of photocatalysis when subjected to the influence of internal electric fields (IEF). Additionally, efforts should be directed towards devising approaches to augment the photocatalytic efficiencies of BiOX-based materials by leveraging the alterations in IEF.

In the realm of water splitting by photocatalysis, the conduction positions of the catalysts belonging to the BiOX series do not exhibit dominance. Therefore, it is imperative to undertake further investigation into effective strategies aimed at enhancing their conduction positions or diminishing the reduction potential.

This can be achieved by activating water molecules through the implementation of modifications to the BiOX catalysts. In future investigations pertaining to the production of hydrogen through photocatalytic water splitting utilizing BiOX nanomaterials, it is imperative to conduct a comprehensive examination of the underlying reaction mechanisms and the advancement of an effective hydrogen production system. Additionally, it is crucial to undertake a study on the durability and long-term viability of BiOX photocatalysts in real-world applications. There exist other challenges that necessitate resolution in order to attain a substantial production of hydrogen and fulfill practical utilization. These challenges encompass the concurrent generation of hydrogen and oxygen, the segregation of hydrogen and oxygen, and the development of large-scale reactor designs.

According to the ongoing investigations of scientific experts, BiOX ought to be capable of pretty good activities in the degradation of organics and hexavalent chromium ions. Having said that, additional research and development is still required for the relevant industrial application technology or recycling approach. In particular, we need to make an effort to collaborate with some of the more established practices in the business, such as the chemical treatment of contaminants, the biological degradation of pollutants, and other similar practices.

The current capabilities of BiOX in terms of nitrogen fixation and carbon dioxide reduction by a photocatalysis process remain insufficient to fulfill the demands of industrial-scale manufacturing. Hence, it is imperative to undertake additional comprehensive investigation and scholarly inquiry. Furthermore, the implementation of a rational modification strategy is of utmost importance. It is imperative to develop rational modification tactics that are grounded on dependable reaction mechanisms. Hence, it is essential to undertake a comprehensive investigation into the intricate workings of photocatalytic reduction of carbon dioxide and nitrogen fixation in forthcoming research endeavors. Furthermore, it is imperative to incorporate synergistic techniques into the photocatalytic carbon dioxide reduction process, including the utilization of photothermal synergistic catalysis and the exploitation of developing surface frustrated synergies.

The primary emphasis in ongoing research on BiOX in the field of electrocatalytic carbon dioxide reduction is centered on the alteration of the bismuth–oxygen layer or the separation between two neighboring bismuth–oxygen layers. Numerous alternative approaches, such as the integration of metal/non-metal atoms into the bismuth–oxygen layer, the creation of planar strain, the introduction of vacancies in the bismuth–oxygen layer, the construction of vertical/lateral heterostructures, or the formation of moiré superlattices in BiOX, have not yet been utilized for the purpose of modifying BiOX in the context of electrocatalytic carbon dioxide reduction. Despite the fact that the existing electrocatalysts utilizing BiOX as templates exhibit notable faradaic efficiency and selectivity towards carbon dioxide reduction, the process necessitates a substantial applied voltage and mostly yields formate as the end product. These factors impose limitations on the feasibility of industrial implementation. Investigating the efficacy of employing ways to enhance the photocatalytic properties of BiOX in order to decrease the required voltage for carbon dioxide reduction and perhaps alter the resulting product to C_2 or even C_3

compounds is imperative for the advancement and practical implementation of BiOX in industrial settings.

Numerous studies have been conducted on BiOX; however, the research on their use as photocatalysts and electrocatalysts remains separate and distinct. To date, there has been a lack of research dedicated to investigating the dual functionality exhibited by BiOX. Therefore, it is crucial to build well-designed devices that can effectively utilize BiOX as templates for electrocatalysts and photocatalysts. This is essential for the future advancement and industrial implementation of BiOX.

The investigation of the photocatalytic antibacterial properties exhibited by catalysts belonging to the BiOX class holds substantial importance. Photocatalytic technology has been shown to be an efficient means of mitigating antibiotic resistance resulting from the utilization of antibiotics. The recovery and diffusion capacities of the BiOX family catalysts play a critical role as limiting variables in their antibacterial applications. In future research endeavors, it is imperative to integrate BiOX series catalysts with appropriate carriers in order to address the aforementioned challenges.

The assessment of photocatalytic degradation using BiOX catalysts has predominantly been carried out in artificial solutions containing antibiotic doses below the milligram per liter threshold. Typically, the amount of antibiotics in effluents from wastewater treatment plants and surface water is commonly seen to be between the range of nanograms per liter or micrograms per liter. Therefore, it is imperative to prioritize the utilization of BiOX photocatalysts in the practical context of antibiotic breakdown, particularly in scenarios with minimal antibiotic concentrations. The superior oxidation potential of hydroxyl radicals, along with their lower selectivity, has led to the predominant explanation for the reaction mechanism among hydroxyl radicals as well as pollutants being attributed to their adsorption on catalyst surfaces. Prospective investigations should pay attention to the absorbability of BiOX photocatalysts for the removal of antibiotic and the interference of other contaminants. The degradation pathway elucidates the destiny and transformation of antibiotics in the context of the photodegradation process. Therefore, it is imperative to thoroughly investigate the atomic-level photocatalytic degradation mechanism in order to enhance the effectiveness of antibiotic degradation. The presence of anions in actual water matrices has a detrimental effect on the efficacy of antibiotic removal through the process of BiOX-based catalyst photodegradation. Therefore, it is imperative to create catalysts that possess exceptional capacity to mitigate the impact of inorganic salts. In addition to the prevailing concern regarding the use of antibiotics, there is an increasing demand for wastewater treatment facilities to address the issue of mitigating the release of antibiotic-resistant bacteria and genes associated with antibiotic resistance. The development of a novel technical pathway for the concurrent elimination of antibiotics, antibiotic-resistant bacteria, and antibiotic resistance genes through the utilization of BiOX photocatalysts is of utmost importance and requires immediate attention. Therefore, it is imperative to do further investigation into the integration of photocatalysis with other treatment methods, including electrocatalysis, the Fenton process, and biodegradation.

However, it is imperative to consider two crucial concerns with BiOX photocatalysts from a practical standpoint. The objective is to develop and produce BiOX photocatalysts that possess broad visible light absorption, exceptional pollutant mineralization capabilities, and high stability. Additionally, it is important to assess the suitability of these photocatalysts for various types of untreated water sources. Another approach involves the development of optimized photocatalytic reactors capable of achieving high levels of efficiency in mass transfer, light absorption, and the recovery or reutilization of photocatalysts. The resolution of the two challenges pertaining to BiOX photocatalysts plays an important role in elucidating the feasibility of implementing these catalysts on a broad scale for water treatment purposes. Therefore, it is crucial to focus on enhancing visible light absorption, organic mineralization, and BiOX stabilization, as well as optimizing the design of photocatalytic reactors and their associated process parameters. These advancements are necessary to enable the widespread use of BiOX nanomaterials in practical, large-scale applications. These themes should be considered as the fundamental basis for engaging in rigorous research discussions in future investigations.

Photocatalysis represents a viable and ecologically sustainable approach to address the issue of wastewater management. Given the significant photocatalytic activity exhibited by BiOX, there is a growing emphasis on the advancement of BiOX-based photocatalysts within the field of photocatalysis. This pursuit holds considerable potential for driving progress in the realm of environmental protection sectors.

www.ingramcontent.com/pod-product-compliance
Ingram Content Group UK Ltd.
Pitfield, Milton Keynes, MK11 3LW, UK
UKHW051850100926
8556IPUK00005B/548

9 780750 359320